J	MY	11/24/2009
523.8	Croswell, Ken.	
Croswell	The lives of stars /	

WITHDRAWN

19.95

The Lives of Stars

Ken Croswell

Boyds Mills Press
Honesdale, Pennsylvania

Acknowledgments

I thank all those who produced the spectacular images I selected for this book. I especially thank Tony Hallas, who digitally reprocessed nearly all the images to bring out the very best color and sharpness.

I thank Laura Gander for the superb reproduction of the images.

Once again I'm very grateful to Tim Gillner for creating such a bold, colorful, and beautiful design for the book.

I thank James Kaler at the University of Illinois at Urbana-Champaign; Robert Mathieu at the University of Wisconsin at Madison; Richard Pogge at Ohio State University; Steven Stahler at the University of California at Berkeley; and Austin Harker, a fifth-grader in Pagosa Springs, Colorado, for reading and critiquing the entire manuscript. I also thank Jay Holberg at the University of Arizona for information about Sirius and Stan Woosley at the University of California at Santa Cruz for examining the elemental origins given in the nucleosynthesis table.

For their support of this ambitious project, I thank my agent, Russell Galen; my editor, Andy Boyles; and my publisher, Kent L. Brown Jr.

IMAGE CREDITS

Cassiopeia A, front cover, pages 38, 69: Spitzer Space Telescope, Hubble Space Telescope, and Chandra X-Ray Observatory. NASA/Jet Propulsion Laboratory–Caltech/O. Krause (Steward Observatory).

Crab Nebula, back cover, pages 1, 16–17, 36–37, 63: Hubble Space Telescope. NASA, European Space Agency (ESA), and J. Hester and A. Loll (Arizona State University).

Lagoon Nebula, pages 2–3, 11, 65: Kitt Peak 4-Meter Mayall Telescope. National Optical Astronomy Observatories (NOAO)/ Association of Universities for Research in Astronomy (AURA)/National Science Foundation (NSF).

NGC 346 in Small Magellanic Cloud, pages 4–5, 18–19, 54–55: Hubble Space Telescope. NASA, ESA, and A. Nota [ESA/Space Telescope Science Institute (STScI), STScI/AURA].

Galactic Bulge, pages 6–7, 68: Hubble Space Telescope. NASA, ESA, and K. Sahu (STScI).

Orion Nebula, pages 9, 66: Hubble Space Telescope. NASA, ESA, M. Robberto (STScI/ESA), and the Hubble Space Telescope Orion Treasury Project Team.

Horsehead Nebula, pages 10, 64: Very Large Telescope. European Southern Observatory.

Eagle Nebula, pages 12, 65: Hubble Space Telescope. NASA, ESA, STScI, and J. Hester and P. Scowen (Arizona State University).

Cone Nebula, pages 13, 56–57: Hubble Space Telescope. NASA, H. Ford (Johns Hopkins University), G. Illingworth (University of California at Santa Cruz/Lick Observatory), M. Clampin (STScI), G. Hartig (STScI), the Advanced Camera for Surveys Science Team, and ESA.

T Tauri, page 14: Image obtained as part of the Two Micron All Sky Survey (2MASS), a joint project of the University of Massachusetts and the Infrared Processing and Analysis Center/California Institute of Technology, funded by NASA and NSF.

Hertzsprung-Russell Diagram, pages 18–19: Tim Gillner, with data from Ken Croswell.

Veil Nebula, pages 20–21, 39: WIYN 0.9-Meter Telescope at Kitt Peak. T. A. Rector (University of Alaska at Anchorage) and Wisconsin, Indiana, Yale, and NOAO (WIYN)/NOAO/AURA/NSF.

SCR 1845-6357, page 23: European Southern Observatory.

Mira, page 24: Hubble Space Telescope. Margarita Karovska (Harvard–Smithsonian Center for Astrophysics) and NASA.

Ring Nebula, pages 27, 68: Hubble Space Telescope. The Hubble Heritage Team (AURA/STScI/NASA).

Helix Nebula, page 28: Hubble Space Telescope and the NSF's 0.9-Meter Telescope at Kitt Peak. NASA, NOAO, ESA, the Hubble Helix Nebula Team, M. Meixner (STScI), and T. A. Rector (National Radio Astronomy Observatory).

Dumbbell Nebula, pages 29, 63: Very Large Telescope. European Southern Observatory.

Eskimo Nebula, page 30: Hubble Space Telescope. NASA, Andrew Fruchter, and the Early Release Observations Team [Sylvia Baggett (STScI), Richard Hook (Space Telescope European Coordinating Facility (ST-ECF)), Zoltan Levay (STScI)].

Cat's Eye Nebula, page 31: Hubble Space Telescope. J. P. Harrington and K. J. Borkowski (University of Maryland) and NASA.

Sirius, page 32: Hubble Space Telescope. NASA, H. E. Bond and E. Nelan (STScI), M. Barstow and M. Burleigh (University of Leicester, United Kingdom), and J. B. Holberg (University of Arizona).

Orion, page 35: Akira Fujii.

NGC 3370, pages 40, 64: Hubble Space Telescope. NASA, the Hubble Heritage Team, and A. Riess (STScI).

Crab Pulsar, pages 43, 66: Hubble Space Telescope and Chandra X-Ray Observatory. NASA/Hubble Space Telescope/Chandra X-Ray Center/Arizona State University/J. Hester, et al.

Cygnus X-1, pages 44, 62: ESA. Illustration by Martin Kornmesser, ESA/ECF.

Albireo, page 46: Johannes Schedler (Panther Observatory).

SN 1994D in galaxy NGC 4526, pages 49, 69: Hubble Space Telescope. NASA, ESA, the Hubble Key Project Team, and the High-Z Supernova Search Team.

Pleiades, pages 50–51, 67: Palomar 48-Inch Schmidt Telescope. NASA, ESA, and AURA/Caltech.

Quintuplet Cluster, page 52: Hubble Space Telescope. Don Figer (STScI) and NASA.

M80, pages 53, 62–63, 64–65, 66–67, 68–69: Hubble Space Telescope. The Hubble Heritage Team (AURA/STScI/NASA).

Extrasolar Planet Around Red Dwarf, page 59: NASA, ESA, and G. Bacon (STScI).

Earth, page 60: Apollo 17. NASA.

Text copyright © 2009 by Ken Croswell
All rights reserved

Boyds Mills Press, Inc.
815 Church Street
Honesdale, Pennsylvania 18431
Printed in China

The text of this book is set in 12-point Cambria.

10 9 8 7 6 5 4 3 2 1

Library of Congress Cataloging-in-Publication Data

Croswell, Ken.
 The lives of stars / Ken Croswell. —1st ed.
   p. cm.
 Includes index.
 ISBN 978-1-59078-582-9 (hardcover : alk. paper)
 1. Stars—Popular works. I. Title.

QB801.6.C76 2009
523.8—dc22

2008033913

Contents

The Lives of Stars 4
Star Light, Star Bright 6
The Birthplace of Stars 8
A Star Is Born 14
The H-R Diagram 16
Main-Sequence Stars 20
Brown Dwarfs 22
Red Giants 24
Planetary Nebulae 26
White Dwarfs 32
Supergiants and Supernovae 34
Cepheids 40
Neutron Stars and Pulsars 42
Black Holes 44
Double Stars 46
When Little Stars Explode 48
Star Clusters 50
Origin of the Elements 54
Extrasolar Planets 58
Life in Space? 60

Glossary 62
Index 70

The Lives of Stars

Like people, stars are born, live, and die.

On a dark night, you can see thousands of stars. All these stars belong to our Galaxy, the Milky Way.

Without stars, you could not live. Before the Earth was born, stars made the oxygen you breathe, the calcium in your bones, and the iron in your blood.

And one star—the Sun—shines on the Earth and keeps it warm.

That's right, the Sun is a star. In fact, the Sun is the closest star of all. So the Sun looks bigger and brighter than any other star.

A star is a hot, glowing ball in space. It makes its own visible light. A planet doesn't. Instead, a planet reflects the light of the star it goes around. The Earth is a planet that orbits the Sun.

Our Galaxy has more than 100,000,000,000 stars. That's a lot of stars!

Like snowflakes, however, no two stars are the same.

Some stars shine brightly. If the Sun were one of these stars, the Earth would get so hot its oceans would evaporate.

Some stars shine dimly. If the Sun were one of these stars, noon would be darker than a moonlit night, and Earth's oceans would freeze.

Some stars are dead: black holes that emit no light at all.

Some stars are huge. If the Sun were as big as the biggest star, it would swallow Mercury, Venus, Earth, Mars, and Jupiter—the five planets closest to the Sun.

Some stars are tiny. They are as small as a city on Earth.

Some stars are blue. Some stars are white. Some stars, like the Sun, are yellow. Some stars are orange. And some stars are red.

Some stars are nearly as old as the universe. But some stars are younger than you.

Like people, stars are born, they live, and they die. But stars live much longer than people. Stars live for millions, billions, or trillions of years.

This book shows how stars live and die.

Stars are born in beautiful clouds of gas and dust.

Then the stars shine, lighting and warming the planets that go around them. Some stars may even support life on those planets.

Then the stars die. Some stars explode. Most stars, however, puff off their outer layers.

Either way, a dying star hurls new elements—such as oxygen, calcium, and iron—into space. So when a star dies, it can give off elements that life needs.

Star Light, Star Bright

The hotter and bigger a star, the brighter it shines.

Stars shine because they are hot. The hotter and larger a star, the more light it sends into space.

Some stars look brighter than others. But that doesn't mean they give off more light.

The brightest star at night is Sirius, pronounced just like the word *serious*. Of course, the Sun looks even brighter than Sirius. Does that mean the Sun emits more light?

No. The Sun is much closer to Earth than Sirius is. To find out which star emits more light, we need to know how far the Sun and Sirius are.

Astronomers can measure a star's distance. Here's how. Because the Earth goes around the Sun, the Earth is on the opposite side of the Sun in summer than in winter. As a result, in each season we view Sirius from a slightly different angle, which causes the star's position in the sky to shift slightly. This shift is called parallax. The larger a star's parallax, the closer the star is to Earth.

To understand parallax, stand in your home and look outside at a nearby object, such as a tree, that's in front of you. Then walk to the left and walk to the right. The nearby tree will seem to shift back and forth. It will shift more than a distant tree will. In the same way, as the Earth goes around the Sun, a nearby star seems to move more than a far-off star.

Sirius is in the southern sky. From 1881 to 1883, two astronomers in South Africa, David Gill and William Elkin, used a telescope to measure the parallax of Sirius. They found that the star is 8.6 light-years away. That means light takes 8.6 years to travel from Sirius to Earth. So when you see Sirius, you see the star as it was 8.6 years ago.

In contrast, the Sun is so close—about 93 million (93,000,000) miles away—that its light takes only 8 *minutes* to reach the Earth.

Light is fast. It moves at 671 million miles per hour. So a light-year—the distance light travels in a year—is huge: about 5.88 *trillion* (5,880,000,000,000) miles.

To see how large a light-year is, imagine a big map, with the Sun and Earth one inch apart. On this map, Jupiter is five inches from the Sun. Distant Pluto is forty inches from the Sun. On the same map, a light-year is one *mile*; so Sirius is 8.6 miles away.

Once astronomers know the distance of Sirius, they can use how bright it looks to calculate that it emits 22 times more light than the Sun. If we could view both Sirius and the Sun from the same distance, Sirius would look 22 times brighter. We say the luminosity of Sirius—the total amount of light it emits into space—is 22 times greater than the Sun's luminosity.

So just because a star looks brighter does not mean it emits more light. You also have to know

its distance to figure out how luminous it is.

Stars shine because they are hot—much hotter than the hottest desert on Earth.

Anything that is very hot gives off light. For example, on an electric stove, the coils get hot and glow. And the embers of a fire glow because they are hot.

The Sun's surface is 9,940 degrees Fahrenheit, or 5,780 Kelvin. (Kelvin is a temperature scale that astronomers use.)

Some stars, such as Sirius, are even hotter. The hotter a star, the more light every square inch of its surface sends into space.

To calculate how a star's surface temperature affects its luminosity, multiply the temperature four times. For example, if the Sun's surface were twice as hot as it is, every square inch would emit 2 x 2 x 2 x 2 = 16 times more light. If the Sun's surface temperature were three times what it is, then every square inch would give off 3 x 3 x 3 x 3 = 81 times more light. And if the Sun's temperature were half what it now is, then every square inch of the Sun's surface would emit 1/2 x 1/2 x 1/2 x 1/2 = 1/16 as much light.

So a small rise in a star's surface temperature brightens the star a lot. And a small drop in surface temperature dims the star a lot.

This rule is called the Stefan-Boltzmann law. It is named for Josef Stefan and Ludwig Boltzmann, two physicists in Austria. They discovered this rule in the late 1800s.

A star's size also affects its luminosity. The bigger a star, the more square inches its surface has—so the more light it casts into space.

The Sun's diameter is 864,900 miles. That's more than a hundred times the Earth's diameter. The diameter is the length of a straight line that passes from one side of a star, through the center, to the other side.

To calculate how size affects luminosity, just multiply a star's diameter by itself. For example, if the Sun's surface temperature stayed the same but its diameter doubled, the Sun would emit 2 x 2 = 4 times more light. That's because its surface would then have four times as many square inches, all emitting light.

If the Sun's diameter tripled, it would emit 3 x 3 = 9 times more light. And if the Sun shrank to half its current size, then it would emit 1/2 x 1/2 = 1/4 as much light—if its surface temperature stayed the same.

Sirius is hotter and larger than the Sun. That's why it emits more light than the Sun.

Although stars are hot, they are born in the cold of space—in dark clouds of gas and dust called nebulae.

The Birthplace of Stars

Stars are born in clouds of gas and dust.

The space between the stars is almost empty. But this space has beautiful clouds. Unlike the clouds in Earth's sky, space clouds aren't made of water. They're made of gas and dust. Some of them give birth to new stars.

A cloud in space is called a nebula. That's the Latin word for *cloud*. The plural is *nebulae*.

The most famous nebula is the Orion Nebula. It is 1,350 light-years from Earth. If you know where to look—south of Orion's three-star belt—you can see the Orion Nebula without using a telescope. It looks like a misty cloud.

Nebulae have different shapes. The Horsehead Nebula looks like a knight in a chess game. The Lagoon Nebula looks like a lagoon.

The fancy name for the space between the stars is "the interstellar medium." That's because *inter* means "between." So *interstellar* means "between the stars."

The interstellar medium is emptier than anything on Earth. To see how empty, imagine a small cube. Each side of this cube is only one centimeter long. So the cube is about as small as a sugar cube.

Imagine a cube this small filled with some of the air you breathe. The cube would have 25 quintillion (25,000,000,000,000,000,000) molecules in it. So we say the density of air is 25 quintillion molecules per cubic centimeter.

In contrast, the interstellar medium on average has only one atom per cubic centimeter. That means if you put the cube in interstellar space, the cube would probably have just one atom in it.

Even the densest interstellar material—such as the beautiful Horsehead Nebula—is nearly empty. It has only a few tens of thousands of molecules per cubic centimeter. That's still much, much emptier than the air you breathe.

Yet our Galaxy is so big that all this interstellar matter adds up. Altogether, the Milky Way's interstellar matter has 5 to 10 billion times more mass than the Sun.

Of this interstellar material, 99 percent is gas and 1 percent is dust. The dust makes some clouds, such as the Horsehead Nebula, look dark, because dust blocks starlight. If the Earth's air were 1 percent dust, the air would be so dark that you wouldn't be able to see your own feet.

Nebulae are made mostly of hydrogen, the lightest and most common element. In space, hydrogen comes in three different forms.

Most interstellar hydrogen gas is made of separate hydrogen atoms. Each hydrogen atom

Horsehead Nebula

has one proton, a particle with positive electric charge. Near the proton is one electron, which has negative electric charge. Opposite charges attract each other, so the proton attracts the electron, and they stay close together. Each hydrogen atom is neutral—it has no charge because the two charges cancel each other. Astronomers therefore call it neutral atomic hydrogen gas. It is also called H I (pronounced "H one").

Neutral atomic hydrogen gas emits radio waves that are 8 inches, or 21 centimeters, long. Radio waves travel as fast as light, but their wavelengths are much longer. Unlike visible light, radio waves go through dust. So astronomers can detect radio waves throughout the Galaxy.

To detect radio waves, astronomers use special telescopes called radio telescopes. Using these telescopes, astronomers detect 21-centimeter radio waves from neutral atomic hydrogen gas. They find that most of the Milky Way's neutral atomic hydrogen gas is in the outer part of our Galaxy.

If a nebula has a very hot star—whose

Lagoon Nebula

surface temperature is more than 50,000 degrees Fahrenheit, or 28,000 Kelvin—its light tears hydrogen atoms apart. In fact, just one hot star can tear hydrogen atoms apart for dozens or even hundreds of light-years in all directions.

Here's how. A hot star emits ultraviolet radiation. Ultraviolet travels at light speed. But it has a shorter wavelength than visible light, so we can't see it. A particle of ultraviolet light has more energy than a particle of visible light. Particles with the shortest ultraviolet wavelengths have so much energy that they tear electrons from protons. These torn-apart hydrogen atoms are called ionized hydrogen, or H II ("H two").

Most H II regions are red. For example, the Lagoon Nebula is red. The reason: when torn-apart hydrogen atoms get back together—when the electrons come near the protons again—they can give off red light.

In the densest and coldest nebulae, something else happens. Two hydrogen atoms join to form a hydrogen molecule: H_2. Molecular gas is so dense it can give birth to new stars.

Unfortunately, hydrogen molecules don't

Eagle Nebula

Color is not true.

usually emit visible light or radio waves. That means astronomers can't easily see them. So astronomers observe another interstellar molecule: carbon monoxide, which is made of one carbon atom joined to one oxygen atom.

On Earth, carbon monoxide is a poisonous gas, produced by cars and cigarettes. But in space, carbon monoxide helps astronomers see where molecular gas is. Carbon monoxide emits radio waves that are 2.6 millimeters—about a tenth of an inch—long. Astronomers find that most molecular hydrogen gas is in the inner part of our Galaxy.

The Orion Nebula has molecular gas. So do the Lagoon and Horsehead nebulae as well as the Eagle and Cone nebulae. All these nebulae have newborn stars. The Orion Nebula has thousands of newborn stars.

By interstellar standards—but not by Earth standards!—molecular gas is dense. Thus, a cloud of molecular gas can collapse under its own weight. When that happens, a star is born.

A Star Is Born

Each star owes its life to gravity.

T Tauri

Inside dark nebulae, giant clumps of gas and dust collapse to create new stars. Every year our Galaxy gives birth to about ten new stars. Each star owes its life to gravity.

Gravity is the force that attracts one mass to another. Gravity keeps you from floating into space. That's because the Earth's gravity attracts you to the Earth. The Earth's gravity also keeps the Moon going around the Earth. And the Sun's gravity keeps the Earth going around the Sun.

Gravity can make things move. Drop a pebble and gravity makes it fall, because the Earth's gravity pulls the pebble down.

The same thing happens in space. Inside a nebula, in a clump of gas and dust trillions of miles across, gravity tries to pull the outer parts toward the center, to make the clump collapse.

But a cloud's own gas fights the collapse. Gas pushes outward. This "gas pressure" opposes gravity. If you've ever squeezed a balloon and felt the balloon press back, you've felt gas pressure.

The colder the cloud, the less gas pressure it has. Interstellar clouds can be very cold. Many are just 10 Kelvin, or −440 degrees Fahrenheit. That's colder than Pluto. So their gas pressure is weak, and gravity can overwhelm the gas pressure, making the cloud collapse.

As gravity pulls gas and dust from the cloud's edge inward, they speed up. They crash into the cloud's center, heating it up. If the cloud has lots of mass, lots of gas and dust crash into the center, raising its temperature.

So the cloud's center, which was once cold, gets warm, then hot, then very hot. And remember, anything that is very hot glows. So the center of the collapsing cloud starts to shine—like a star. Indeed, it now *is* a star, for it is creating its own light. A star has been born!

However, astronomers may not see the newborn star's visible light. The star may still be buried in the dusty cloud that gave it birth. The dust blocks the star's visible light.

But the light heats the dust. So the dust starts to glow. It does not emit visible light. Instead, it emits infrared radiation.

Infrared travels at the speed of light. But its wavelength is longer, so our eyes can't see it. Fortunately, unlike visible light, infrared goes through dust. And astronomers can detect infrared, so they can see newborn stars still wrapped in thick dust.

As a newborn star shines, its light carries energy from the star's surface into space. So the star loses heat. But as gravity squeezes the star, the star keeps getting smaller; so gravity heats the star further and its temperature rises more.

Think about that. As a newborn star *loses* heat, it gets *hotter*! Normally, when something loses heat, it gets colder. If you take a hot loaf of bread out of the oven, the bread loses heat and cools down. Likewise, if you pluck a glowing coal from a fire, the coal loses heat and cools down. Because of gravity, a newborn star is the only thing that loses heat yet gets hotter!

The most famous newborn star is 480 light-years away in the constellation Taurus. Its name is T Tauri. T Tauri is still wrapped in the gas and dust that gave it birth. The star shines because gravity is squeezing it and heating it.

T Tauri is only 1 million years old. It probably looks much as the Sun did when it was born, 4.6 billion years ago. T Tauri is to the Sun what a four-day-old infant is to a fifty-year-old adult.

For many millions of years after its birth, a star like T Tauri continues to shrink. As gravity heats it, its center gets hotter and hotter. Eventually, in a star with the Sun's mass, the center reaches 22 million degrees Fahrenheit, or 12 million Kelvin.

Then protons at the star's center join other protons. A proton is the part of an atom with positive electric charge. Normally, two protons push each other away, because they both have positive charge. In the same way, you can't push together the wrong ends of two magnets.

But as the star's center heats up, protons move faster and faster. They move so fast that they crash into one another. When they are so close to each other, a powerful force called the strong force makes them join together, even though they have the same electric charge.

A proton is the center, or nucleus, of a hydrogen atom. So these proton-proton collisions are *nuclear* reactions. They release lots of energy. Right now, nuclear reactions at the Sun's center change hydrogen, the lightest element, into helium, the second lightest element.

Any star, like the Sun, whose center changes hydrogen into helium is called a main-sequence star. Before this stage, it was a pre-main-sequence star, like T Tauri.

Astronomers calculate that a star with the Sun's mass takes 50 million years after its birth to stop shrinking. At that point, the amount of energy the star makes from its nuclear reactions equals the amount it loses by sending light into space from its surface. Also, the star's gas is so hot that it pushes outward with just the right force to hold the star up against the inward force of its own gravity. So the star is now stable.

In addition, particles of light called photons help hold the star up. As photons stream out of the star's center, they push outward, too.

For the Sun, however, this photon pressure is small. But extremely luminous stars have so many photons that photon pressure supplies *most* of the outward force that keeps the star from collapsing in on itself.

Stars shine because they are hot. Why are stars hot? Now you know there are two reasons. First, when a star is born, gravity heats the star. Then, in most stars, gravity raises the central temperature enough to spark nuclear reactions. These nuclear reactions make the energy that keeps the star hot.

So gravity starts a star's heat. Nuclear reactions keep the star hot. And heat and photons hold the star up against the gravity that gave it birth.

Strangely, despite its brilliance, a star doesn't produce energy very well. Believe it or not, your body produces thousands of times more energy per pound than the Sun does—even though the Sun uses nuclear reactions and you don't. It takes a *lot* of mass to force a star to shine.

A star's surface is much cooler than its center. Whereas the center is millions of degrees hot, the surface is only thousands of degrees hot. Still, surface temperature is a key property of any star. In fact, when astronomers talk about a star's temperature, they usually mean its surface temperature, not its central temperature.

A second key property of a star is its luminosity, the amount of light the star emits into space. These two properties, luminosity and surface temperature, let astronomers plot a star on a diagram that tells them even more about its life and eventual death.

The H-R Diagram

It reveals a star's life story.

A century ago, two astronomers—one in Denmark, the other in America—made a big discovery. They did not look through a telescope. Instead, they plotted points on a graph.

Each point stood for a star. Each point's position on the graph stood not for the star's place in space but for its luminosity and surface temperature.

The first astronomer was Ejnar Hertzsprung. He plotted the graph in 1911. Two years later, in 1913, the other astronomer, Henry Norris Russell, made his own plot. He did not know about Hertzsprung's work.

The Hertzsprung-Russell diagram reveals a lot about stars. It separates stars into three groups: main-sequence stars, such as the Sun; giants and supergiants; and white dwarfs.

You can see a modern H-R diagram on pages 18–19. Every point stands for a real star.

To understand the H-R diagram, let's start by finding the star that's most important to us: the Sun. It's a yellow dot near the center of the diagram.

On the H-R diagram, all stars higher than the Sun—whether blue, white, yellow, orange, or red—emit more light than the Sun. So their luminosities are greater than the Sun's. And all stars lower than the Sun—no matter their color—emit less light than the Sun.

Figuring out a star's luminosity requires that astronomers know how bright a star looks and how far it is from Earth.

To say how bright a star looks, astronomers use a number they call apparent magnitude. Many of the brightest stars in the sky are first magnitude, which means their apparent magnitude is around +1. First-magnitude stars look brighter than second-magnitude stars, whose apparent magnitude is around +2, just as first class is better than second class. In turn, third-magnitude stars are fainter than second-magnitude stars.

The faintest stars you can see without using binoculars or a telescope have apparent magnitude +6. But telescopes can see to apparent magnitude +10, to apparent magnitude +20, even to apparent magnitude +30. That's billions of times fainter than your eye can see. So you see why astronomers like telescopes!

A few stars, such as Vega, look so bright that their apparent magnitudes are around 0. And some stars look even brighter. Their apparent magnitudes are negative. For example, Sirius, the brightest star in the night, has an apparent magnitude of –1.44. And the Sun's apparent magnitude is an incredible –26.74.

Apparent magnitude says only how bright a star *looks*. But to say how bright a star *really* is—to give its luminosity, the total amount of light it emits into space—astronomers use *absolute* magnitude. That's what is on the H-R diagram's vertical axis.

To compute a star's absolute magnitude, astronomers must know the star's apparent magnitude and its distance. Then they imagine seeing the star from a standard distance of 32.6 light-years.

For example, if the Sun were 32.6 light-years away, it would look a lot fainter. You could barely see it. It would be apparent magnitude +4.83. So we say the Sun's absolute magnitude is +4.83. And that's where the Sun is on the H-R diagram.

Even Sirius would not be so special. Sirius is 8.6 light-years away. But if we pushed Sirius out to the standard distance of 32.6 light-years, it would look fainter. Rather than having a negative apparent magnitude, it would have a positive apparent magnitude: +1.45. So this is the absolute magnitude of Sirius.

A few rare stars are so luminous they have negative absolute magnitudes. These stars emit more light than Sirius or the Sun.

For example, the blue star Rigel, in Orion, looks bright at night. But it shines from a distance of roughly 900 light-years—about a hundred times farther than Sirius. If we pulled Rigel in and put it just 32.6 light-years from Earth, it would look brighter than anything in the sky except the Sun and the Moon. Rigel's absolute magnitude is −7.4.

Rigel emits 80,000 times more light than the Sun. Each minute, Rigel emits more light than the Sun gives off in a whole month. Rigel is near the top of the H-R diagram.

Most stars, though—95 percent of them—are less luminous than the Sun. They emit only a fraction of the Sun's light. So the Sun is *not* an average star. The least luminous stars are near the bottom of the H-R diagram.

The H-R diagram shows more than just the luminosities of the stars. It also shows their colors.

Color indicates a star's surface temperature. Blue stars are the hottest. They're on the left of the H-R diagram. White stars are also hot, but not as hot as blue stars. Yellow stars, like the Sun, are warm. The coolest stars here are orange and red, which are on the right of the H-R diagram.

To understand this color-temperature pattern, imagine sticking a metal rod into a fire. As the rod heats up, it first glows red, the coolest color. Then, as it gets hotter, it glows orange, then yellow, then white, then blue-white. So red is the coolest color and blue is the hottest.

It's hard to see these star colors. Unlike the colors of a traffic light, star colors are not obvious. The best example is in Orion, which has a bright blue star, Rigel, and a bright red star, Betelgeuse. If you stare at Rigel for thirty seconds, then quickly look at Betelgeuse, you can see they are different colors.

Astronomers can tell a star's surface temperature by observing its color. They can also measure its temperature from its spectrum. A spectrum is a rainbow of starlight. It arranges all the colors a star is emitting by order of wavelength: purple light has the shortest wavelength, then blue light, then green, then yellow, then orange, and finally red, which has the longest wavelength.

Different elements remove different wavelengths of light, thereby imprinting themselves on the star's spectrum. By looking at the pattern of missing colors in a star's spectrum, astronomers classify its spectral type—and deduce its temperature.

Most stars are made of similar elements. But the surface temperature affects which substances appear in a star's spectrum. Thus, the spectrum reveals the star's surface temperature. In a similar way, if it's raining outside, you can tell it's warmer than if it's snowing.

For example, most stars contain helium, but strong spectral lines of helium appear in only the hottest stars, which are blue. Most stars also have calcium, but calcium spectral lines are strong only in cooler stars: yellow stars, such as the Sun, as well as orange and red stars.

There are seven main spectral types. From hot and blue to cool and red, they are O (blue stars), B (also blue), A (white), F (yellow-white), G (yellow, like our Sun), K (orange), and M (red). You can remember this lineup by using "**O**h, **B**e **A** **F**ine **G**uy/**G**irl, **K**iss **M**e!" Then the first letter of each word matches a spectral type.

Just as you can read a star's luminosity from the H-R diagram, so you can read its spectral type. You can see that the Sun is spectral type G. The blue star Rigel is spectral type B. And the red star Betelgeuse is spectral type M.

The amazing thing about the H-R diagram is that it reveals three groups of stars. The most important is a diagonal band that runs from the top left (luminous and blue) to the middle (where the Sun is) to the bottom right (dim and red). Most stars are in this band, so astronomers call it the main sequence.

Now look at the stars in the upper right of the H-R diagram (luminous and red). These stars are called red giants and supergiants. They are cooler than the Sun. But they are very large, so they emit much more light than the Sun.

Now look at the line of stars below the main sequence. Although these stars span all colors from blue to red, astronomers call these stars white dwarfs. These are dying stars. All are much smaller and dimmer than the Sun.

Each group—main-sequence stars, giants and supergiants, and white dwarfs—is a different stage in the life of a star. Of these three stages, the main sequence is the most important. It's the stage our Sun is in.

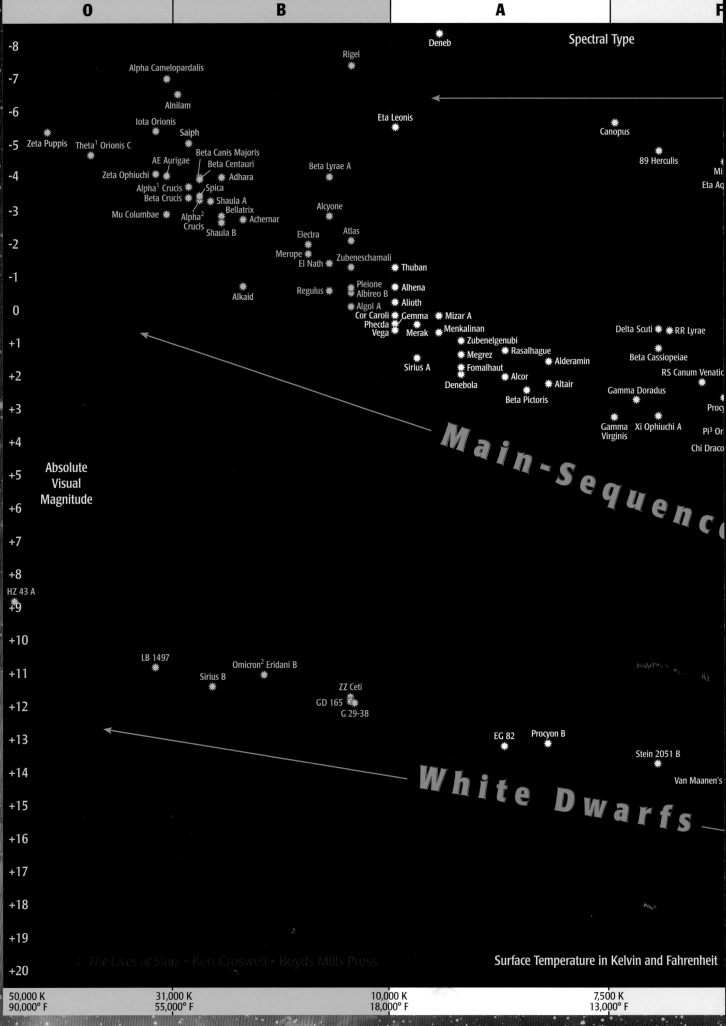

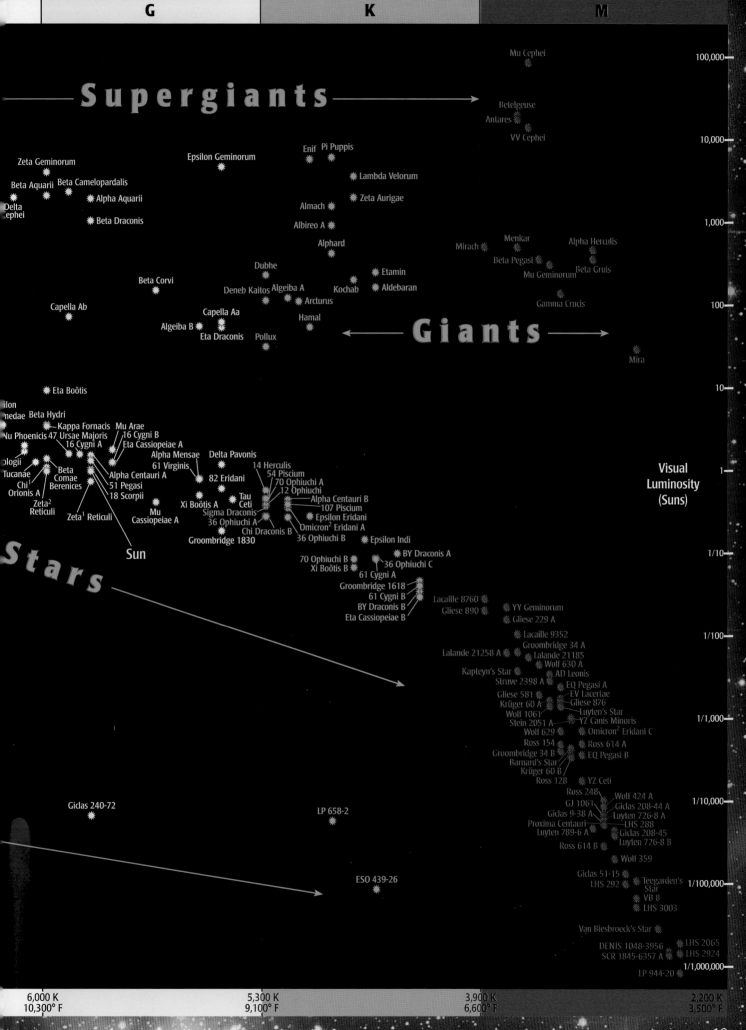

Main-Sequence Stars

For stars, the main sequence is like adulthood.

Main-sequence stars are the most important stars. Our Sun is a main-sequence star. So are most of its neighbors. In fact, 95 percent of all stars are on the main sequence.

For a star, the main sequence is like being an adult. Most main-sequence stars are stable, and most spend a long time in this stage. So they have time to give rise to life on planets that go around them, just as most adults have time to raise children.

The main sequence is the diagonal band of stars that slopes from upper left to lower right on the H-R diagram (see pages 18–19). It includes all kinds of stars—from blue and very bright; to yellow and bright, like our Sun; to red and dim.

The hottest and most luminous main-sequence stars are blue stars of spectral types O and B. For example, blue B-type Regulus, the brightest star in the constellation Leo, is a main-sequence star. You can find Regulus in the blue part of the main sequence on the H-R diagram.

White A-type main-sequence stars shine in many constellations. Sirius—the brightest star of the night—is an A-type main-sequence star. So are Vega and Altair. All these stars are hotter and more luminous than the Sun.

The yellow-white F main-sequence stars are slightly warmer and brighter than the Sun. The closest F-type star is Procyon. It has just started to leave the main sequence.

The Sun is a yellow G-type main-sequence star. So are two of its nearest neighbors, Alpha Centauri A and Tau Ceti.

The orange part of the main sequence includes the K-type stars Alpha Centauri B, Epsilon Eridani, Epsilon Indi, and 61 Cygni. These stars are all cooler and dimmer than the Sun.

But the faintest and coolest main-sequence stars, at the lower right of the H-R diagram, are red. Astronomers call these stars red dwarfs. The typical red dwarf emits less than 1/100 as much visible light as the Sun. Most are spectral type M.

Three fourths of all stars are red dwarfs. Yet they are so dim that you can't see a single one without using binoculars or a telescope.

Most of our neighbors are red dwarfs. Proxima Centauri, 4.24 light-years from the Sun; Barnard's Star, 5.98 light-years from the Sun; Wolf 359, which is 7.8 light-years from the Sun; and Lalande 21185, which is 8.3 light-years from the Sun, are all red dwarfs.

The main sequence includes all sorts of stars: bright blue stars, like Regulus; warm yellow stars, like the Sun; and dim red dwarfs, like Proxima Centauri. What do all these stars have in common—besides lying on a diagonal band of the H-R diagram?

Astronomers wondered.

Then they calculated how stars make energy. They discovered a simple but startling answer: every main-sequence star—whether big or small, hot or cool, blue or red, bright or faint—makes energy the same way: by changing hydrogen into helium at its center.

Furthermore, when astronomers measured the masses of the stars, they discovered something else. They found that the hottest and most luminous main-sequence stars have the most mass.

Measuring a star's mass is hard. But astronomers can sometimes do it if the star is double.

As an example, consider Sirius. It is really two stars. The one you see at night is a hot, A-type main-sequence star that emits 22 times more visible light than the Sun. But a tiny white dwarf star goes around the main star. We call the bright star Sirius A and the dim star Sirius B.

Gravity holds the two stars close to each

Main-Sequence Stars

Spectral Type	Color	Temperature (Kelvin)	Mass (Suns)	Visible Light (Suns)	Approximate Lifetime (years)	Examples
O	Blue	31,000 to 50,000	16 to 100	4,000 to 15,000	3 to 30 million	Zeta Ophiuchi; Mu Columbae
B	Blue	10,000 to 31,000	2.5 to 16	50 to 4,000	30 to 400 million	Regulus; Achernar; Alkaid
A	White	7,500 to 10,000	1.6 to 2.5	8 to 50	400 million to 2 billion	Sirius A; Vega; Altair; Fomalhaut
F	Yellow-white	6,000 to 7,500	1.1 to 1.6	1.8 to 8	2 to 8 billion	Gamma Virginis; Chi Draconis
G	Yellow	5,300 to 6,000	0.9 to 1.1	0.4 to 1.8	8 to 16 billion	Sun; Alpha Centauri A; Tau Ceti
K	Orange	3,900 to 5,300	0.6 to 0.9	0.02 to 0.4	16 to 80 billion	Epsilon Eridani; Epsilon Indi
M	Red	2,200 to 3,900	0.08 to 0.6	0.000001 to 0.02	80 billion to trillions	Proxima Centauri; Barnard's Star

other. Gravity comes from mass. So the more mass the stars have, the more gravity they have.

On average, Sirius A and Sirius B are as far apart as Uranus is from the Sun. The Sun's gravity makes Uranus go around the Sun every 84 years. But by watching Sirius A and B through telescopes, astronomers saw that the two stars take only 50 years to go around each other. Therefore, Sirius A and B together must have more mass than the Sun—because their greater gravity forces them to go around each other faster than Uranus goes around the Sun, even though their distance from each other is the same. When astronomers work out the numbers, they find that Sirius A has about twice the mass of the Sun.

Using this method, astronomers have measured the masses of other stars. They find the same pattern: the hotter, bluer, and brighter a main-sequence star, the more mass it has.

For example, an O-type main-sequence star has more mass than a B-type main-sequence star, which has more mass than an A-type main-sequence star, and so on. The least massive main-sequence stars are the red dwarfs, which are also the coolest and faintest.

Why? It's because of what happens at a star's center. The more mass a star is born with, the more gravity heats its center. And the hotter a star's center, the faster protons move to make helium, so the star makes more energy.

So: the more mass a main-sequence star has,
1. the bigger it is;
2. the hotter it is;
3. the bluer it is; and
4. the more luminous it is.

As a result, just by looking at the H-R diagram, you can see that the B-type main-sequence star Regulus has more mass than A-type Sirius, which has more mass than G-type Alpha Centauri A, which has more mass than K-type Alpha Centauri B, which has more mass than M-type Proxima Centauri.

These rules work *only* for main-sequence stars. They don't work for giants, supergiants, or white dwarfs.

There's another rule main-sequence stars obey. It says how long a star lives.

By measuring the ages of meteorites, scientists find that the Sun, the Earth, and the solar system have existed for 4.6 billion (4,600,000,000) years. Astronomers calculate that the Sun has enough fuel to live for another 7.8 billion years. So the Sun's total lifetime is 12.4 billion years.

The rule for star lifetimes is simple. But at first, it makes no sense. Here it is: the *more* mass a main-sequence star has, the *faster* it dies.

To a star, mass is fuel. So a star born with lots of mass has more fuel than a star born with little mass.

But all that mass also causes gravity to press on the star's center and heat it. The heat makes the star burn its fuel much faster.

For example, a star with twice the mass of the Sun shines about twenty times more brightly. So even though it has twice the Sun's fuel, it burns that fuel twenty times faster. As a result, the star lives only 1/10 as long as the Sun.

It's like a rich man who spends his money fast. He'll go broke long before a poor man who spends his money wisely.

The least massive main-sequence stars—the red dwarfs—are the "wisest" of all. Some red dwarfs burn their fuel so slowly they will live for more than a trillion years. That's much longer than the Sun.

Some stars, though, are born with even less mass than red dwarfs. They never sustain nuclear reactions. Astronomers call them brown dwarfs.

Brown Dwarfs

Stars too small to burn their fuel still shine.

Brown dwarfs are born with little mass. So they never get hot enough to start and continue the nuclear reactions that power main-sequence stars like the Sun.

But when brown dwarfs are young, they are warm. Thus, they emit light. So astronomers can see these "failed stars." The nearest brown dwarf is just a few light-years from Earth.

You can see a brown dwarf in the painting here. At left, in the background, is a *red* dwarf. At right is a brown dwarf that goes around it. Both stars are 13 light-years from Earth, in the southern constellation Pavo the Peacock.

As you can see, the brown dwarf isn't brown. Some brown dwarfs are reddish, because they are cool. This brown dwarf is dark magenta. That's because the brown dwarf has sodium atoms. They absorb yellow and green light, leaving red and a little blue light to mix together. This mix makes purple or magenta.

Why do astronomers call these stars brown dwarfs when they aren't even brown? Some people think it's not a good name.

However, astronomers were smart about one thing. They predicted brown dwarfs would exist long before anybody saw one.

Here's how they did it. They knew that nebulae give birth to stars with different masses. The most massive stars get very hot and shine bright and blue. Less massive stars don't get as hot, so they are yellow, like the Sun. They shine less brightly. The least massive main-sequence stars are the red dwarfs, which burn their fuel slowly. They are much cooler and fainter than the Sun.

But suppose a nebula gives birth to a star with even less mass than a red dwarf. What happens?

The star might stay so cool that it never sparks nuclear reactions.

Like other newborn stars, this star slowly gets smaller. As gravity heats it, the star warms up. Because of its warmth, the star glows red, like a red dwarf. In fact, the star is even spectral type M.

But the star can shrink only so much; thus, gravity can heat it only so much. Then the star starts to cool, ending any nuclear reactions. The star never gets to be a main-sequence star like the Sun. Instead, the star cools further and thus fades.

In the early 1960s, astronomers calculated that these dark stars would have less than 8 percent of the mass of the Sun. That's about 80 times more mass than Jupiter, the largest planet that goes around the Sun.

Astronomers used their telescopes to try to find these dim stars. But they didn't find any. Brown dwarfs are faint. They're hard to see. Brown dwarfs are brightest when they're young. But it's hard to tell a young brown dwarf from a red dwarf.

Then, in 1995, astronomers finally succeeded. They discovered the first definite brown dwarf. It was orbiting a red dwarf named Gliese 229, which is 19 light-years from Earth.

The astronomers used a telescope with a mask that blocked the glare of the red dwarf. Then they saw a much fainter star beside it. They called the dim star Gliese 229 B.

What was Gliese 229 B? The astronomers used another telescope to find out. This telescope recorded the faint star's spectrum. In the spectrum was something no one had ever seen in a star before: methane.

Methane molecules get torn apart by heat. The

Artist's interpretation

Sun doesn't have any methane, nor does any other normal star, because normal stars are too hot. But Gliese 229 B did have methane—because it's so cool it's a brown dwarf! Its surface temperature is only 1,000 Kelvin. That's about 1/6 the surface temperature of the Sun.

After this discovery, astronomers found many other brown dwarfs. The nearest is a *double* brown dwarf that goes around the orange main-sequence star Epsilon Indi. These brown dwarfs are just 12 light-years from Earth.

After astronomers discovered brown dwarfs, they realized they needed to add two new spectral types. The new spectral types are for stars even cooler than M dwarfs.

The seven main spectral types are O, B, A, F, G, K, and M. Now astronomers have added spectral types L and T.

Some brown dwarfs are spectral type M. But most brown dwarfs are spectral types L or T. T is cooler than L. T stars are so cool they have methane. Both Gliese 229 B and the brown dwarf shown here are spectral type T. As a brown dwarf ages, its spectral type changes from M to L to T.

A typical brown dwarf has a few dozen times more mass than Jupiter—but is slightly *smaller* than Jupiter, because the brown dwarf's greater gravity squeezes the star so much. Thus, a brown dwarf is much denser than Jupiter.

Like the Sun, brown dwarfs can have planets. Imagine a planet going around a nearby brown dwarf. At first the planet receives a dim reddish light from its sun. But then that sun turns dark. The planet grows cold. The only light it gets comes from other stars—like Sirius, Alpha Centauri, and our own Sun.

Red Giants

They're bigger and brighter than the Sun.

Mira

Imagine a bright red star. It is so big it has swallowed one or two of its own planets. And it is so bright it has fried another planet that once had water and life.

Where is this deadly star? It is the Sun—but 7.7 billion years from now.

Today the Sun is a main-sequence star. In the far future, though, the Sun will be a red giant, shining a hundred times more brightly.

Some stars have already become giants. You can find them on the H-R diagram—they have an absolute magnitude around 0. For example, Capella is a pair of yellow G-type giant stars. Capella is the sixth brightest star in the night, lying 43 light-years away.

Arcturus and Aldebaran are orange K-type giant stars. Arcturus, the fourth brightest star in the night, is 37 light-years away. If it replaced the Sun, its surface would reach a third of the way to Mercury, the planet closest to the Sun. Aldebaran, the brightest star in Taurus, is larger. It would extend halfway to Mercury.

Mira and Gamma Crucis are red M-type giant stars. If Mira were at the center of our solar system, it would engulf Mercury, Venus, Earth, and Mars.

By calculating how main-sequence stars burn their fuel, astronomers have figured out that giants were once main-sequence stars. While on the main sequence, some of these stars were blue and spectral type B, like Regulus. Others, with less mass, were white A-type stars, like Vega; yellow-white F-type stars, like Procyon; or yellow G-type stars, like the Sun. K-type and M-type main-sequence stars live so long that none has yet left the main sequence.

On the main sequence, a star's center changes, or "burns," hydrogen, the lightest element, into helium, the second lightest element. That's how the star makes energy.

When a star is born, it is mostly hydrogen. So it has lots of hydrogen fuel.

However, as the star changes hydrogen into helium, its center starts to fill with helium. It takes four hydrogen nuclei to make one helium nucleus. So as the star changes hydrogen into helium, its center has fewer and fewer separate particles.

Each separate particle pushes outward, holding the star up against its own weight. But with fewer separate particles in the center, the outward pressure would decrease. So the core shrinks a little, getting hotter and denser, thereby raising the outward pressure. The higher temperature and density make the center burn its fuel a bit faster. Therefore, a main-sequence star expands and brightens a little as it ages.

Astronomers estimate that the Sun is 40 percent brighter today than when it began its main-sequence stage, 4.5 billion years ago. And our star is expanding about as fast as your fingernails grow. In another billion years, after more of the Sun's core changes from hydrogen to helium, the Sun will shine 10 percent more brightly than it does today. As a result, the Earth will get so warm that the water in Earth's oceans, lakes, and rivers will start to evaporate. Since life needs water, all life on Earth will end.

Fortunately, this disaster won't happen for a long time. And if Earth still has intelligent life, that life can probably build mirrors in space to block some of the sunlight. These mirrors might even reflect the sunlight to Mars, a cold planet, to warm it up.

However, about 6 billion years from now, the Sun's center will be so full of helium that the Sun won't be able to burn hydrogen there anymore. Instead, the Sun will start burning hydrogen in a layer around the center. As this happens, the Sun will brighten and expand.

When a gas expands, it cools. So the Sun's surface will cool. It will turn deeper yellow, then orange, then red. It will keep getting brighter, until it shines much more brightly than it does today.

The Sun will be a red giant.

As the Sun expands, it will swallow Mercury. It may swallow Venus, too. The Earth may or may not survive. But bright sunlight will roast the Earth. At that time, the outer planets will receive more light than they do now. Perhaps life on Earth can move to their moons, which should be a better temperature then. Or maybe Earthlings can move their planet farther from the Sun.

When the Sun first becomes a red giant, the helium at its center doesn't do anything. The helium nuclei don't join together. That's because a helium nucleus has twice the electric charge of a hydrogen nucleus. So two helium nuclei push each other away much more strongly than two hydrogen nuclei do. Therefore, to burn helium, the Sun's center must heat up. Then the helium nuclei will move so fast that they smash into one another.

Today the center of the Sun has a temperature of 16 million Kelvin, or 28 million degrees Fahrenheit—much too cool to fuse helium. But after the Sun becomes a red giant, the core will get hotter, because gravity will squeeze the core and heat it. When the center of the Sun reaches around 100 million Kelvin, helium nuclei will smash together, and the helium will start to burn.

A star burns its helium by changing the helium into carbon and oxygen. In fact, a carbon nucleus is really just three helium nuclei bound together. And an oxygen nucleus is four helium nuclei.

Eventually the red giant Sun will start pulsating—first expanding, then shrinking, then expanding again. The Sun will shine far more brightly than it does now. At its peak, the Sun will emit thousands of times more visible and infrared light than it does today.

The most famous pulsating red giant star is Mira. Its name means "the Wonderful." People noticed it centuries ago. As Mira expands and shrinks, it brightens and fades. When Mira is brightest, you can see it in the constellation Cetus. When Mira is faintest, you can't see it at all—except through a telescope.

By watching Mira, astronomers have found that its pulsation period—from bright to faint and back to bright again—lasts about 330 days. Mira is 300 light-years from Earth.

A red giant such as Mira sheds lots of material into space, like a dandelion going to seed. When the star dies, it creates a beautiful, multicolored gas bubble around itself called a planetary nebula.

Planetary Nebulae

At the center of each is a hot, dying star.

A planetary nebula is a stunning sight. It looks like a multicolored smoke ring. A planetary nebula comes from a dying red giant star that has cast its outer layers into space.

But planetary nebulae are more than just beautiful. Many planetary nebulae have carbon and nitrogen, two elements that life needs.

The most famous planetary nebula, shown here, is the Ring Nebula. It's in the constellation Lyra.

The colors here are real. But if you look through a telescope, you can't see them. That's because nebulae are faint and your eye can't see colors in faint things. Astronomers use instruments that separate different colors, to tell which colors are brightest.

Planetary means "relating to planets." But planetary nebulae aren't planets—just as brown dwarfs aren't brown.

Here's why the name exists. In 1781, German-born English astronomer William Herschel used a telescope to discover the planet Uranus. He saw it was not a star because it was not a sharp point of light. Instead, it looked like a disk.

Later, when Herschel used his telescopes to study nebulae, he saw that some nebulae looked like disks, similar to the planet he had found. So he called them planetary nebulae.

Planetary nebulae come from red giant stars. A red giant is big and bright. But it casts its outer layers into space, exposing its hot core.

The core is hotter than any main-sequence star. Some of these hot cores have surface temperatures of more than 100,000 Kelvin—much hotter than the Sun's surface, which is 5,780 Kelvin.

Anything that hot emits not just visible light but also ultraviolet radiation. Ultraviolet radiation travels as fast as light, but it has a shorter wavelength, so you can't see it. Photons of ultraviolet radiation also have more energy than photons of visible light.

Even the Sun emits some ultraviolet light. This ultraviolet light tans your skin and creates vitamin D, which strengthens your bones and teeth and also helps prevent cancer. However, too much ultraviolet light harms your skin.

Because the core of a red giant is much hotter than the Sun, it emits much more ultraviolet radiation. The photons of this radiation have so much energy that they cause atoms in the gas the red giant has cast off to glow at visible wavelengths. This glow is what astronomers see through their telescopes—and what we see in the images here.

In fact, if you look closely, you'll see the hot core at the center of each planetary nebula. That tiny core causes the whole planetary nebula to glow.

And the cores *are* tiny. The one at the center of the Dumbbell Nebula is just six times the diameter of the Earth. That's only 1/20 the diameter of the Sun.

The gas in the planetary nebula moves away from the hot core. So the planetary nebula is expanding. By measuring how fast it expands, astronomers calculate that a planetary nebula lasts only a few tens of thousands of years.

When the planetary nebula is gone, the tiny, hot star at its center still shines. A red giant has died—and a white dwarf has been born.

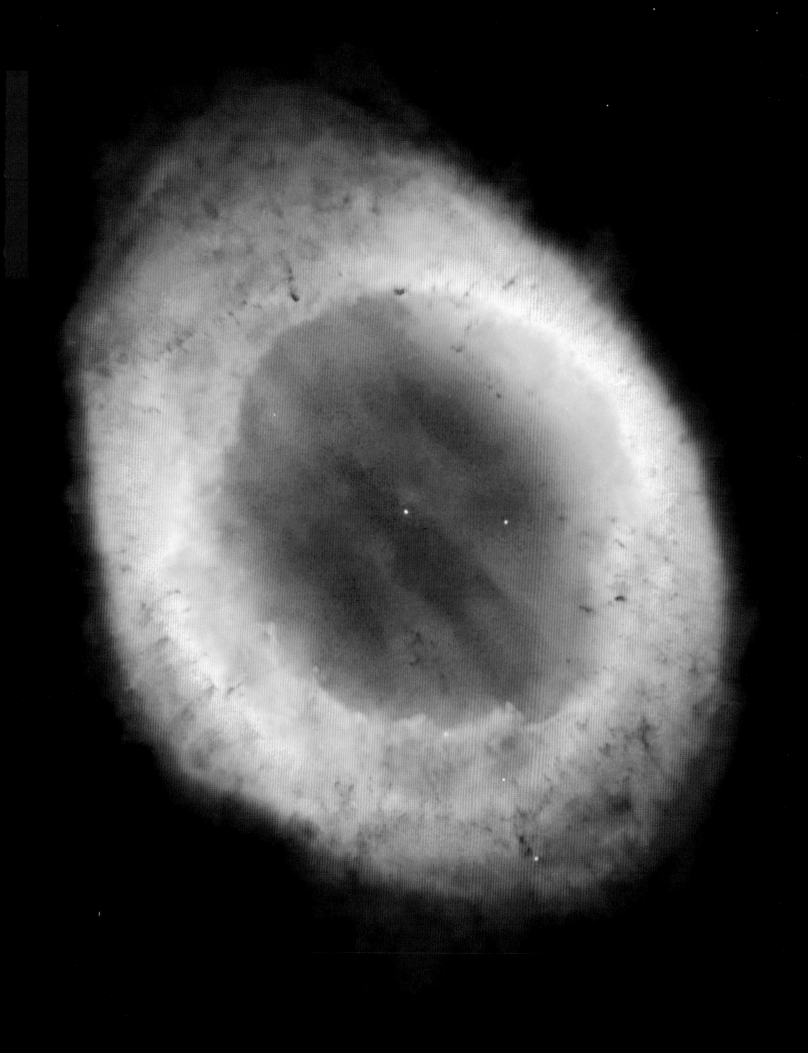

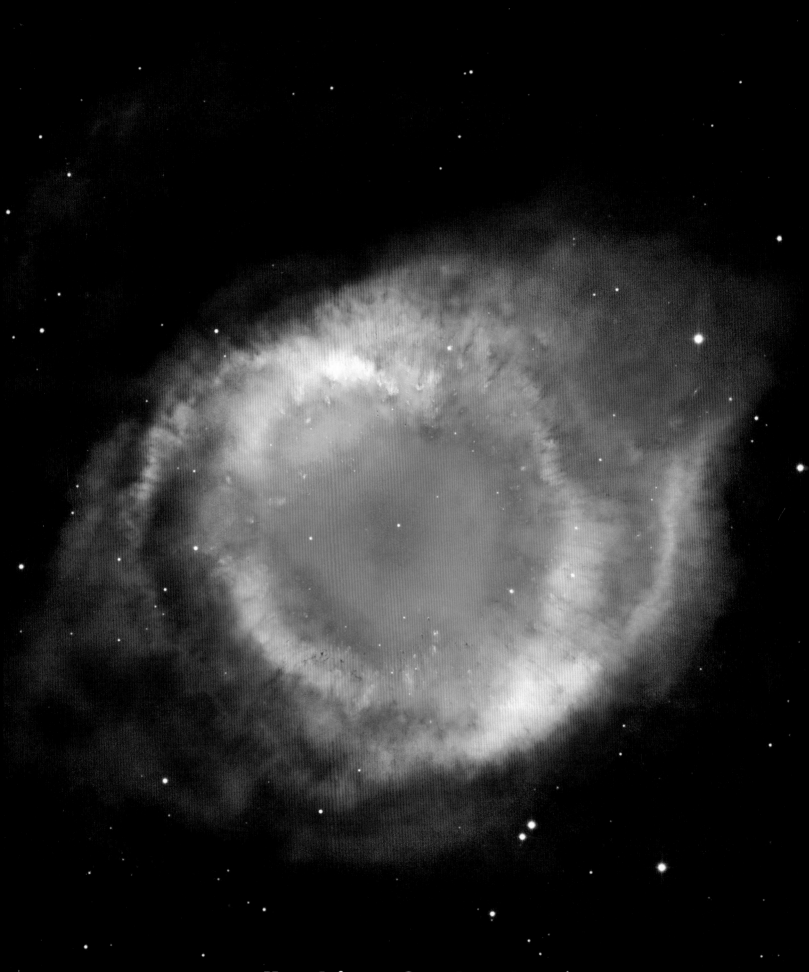

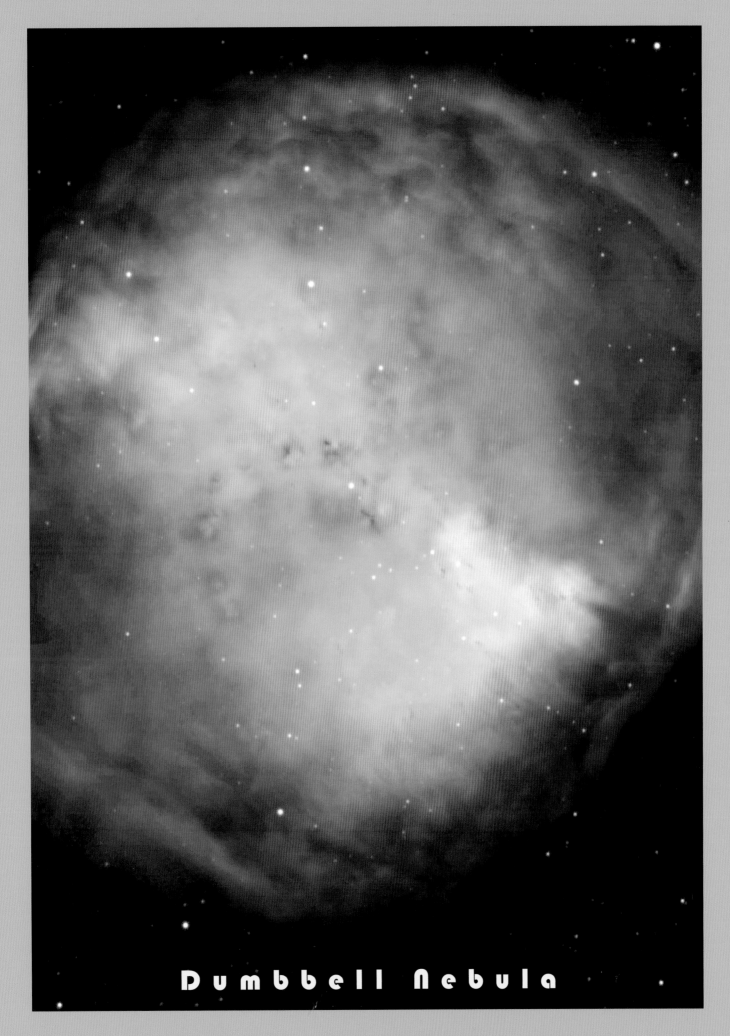

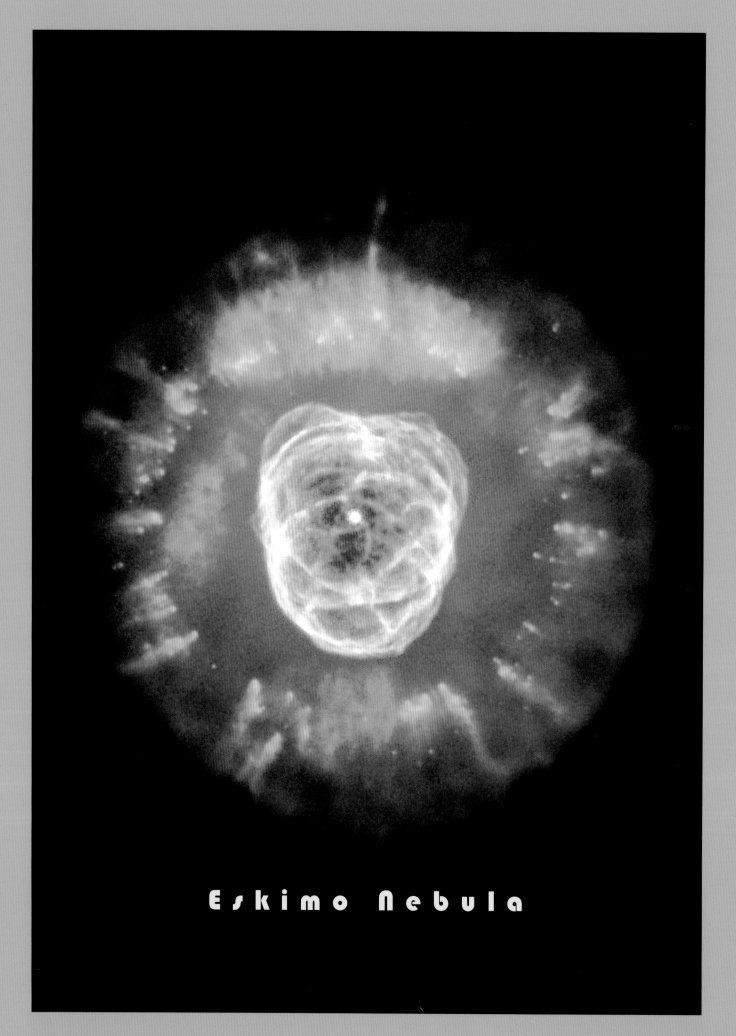

Eskimo Nebula

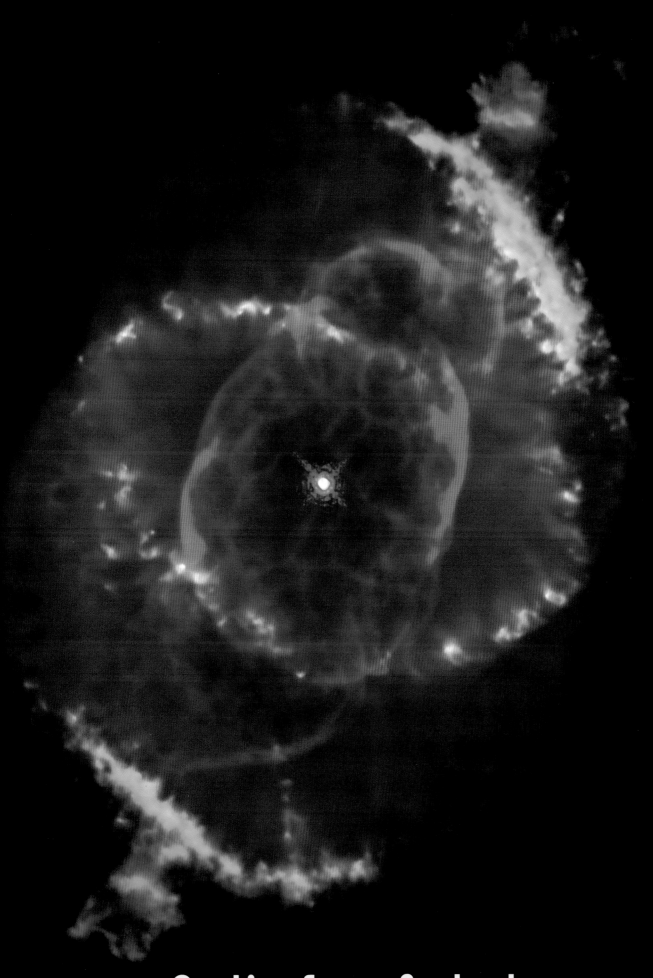

White Dwarfs

A spoonful weighs tons.

White dwarf stars are tiny and faint. They are only about the size of the Earth. But they have about as much mass as the Sun. White dwarfs are so dense that just one spoonful weighs tons.

One star in twenty is a white dwarf. The nearest is just a few light-years away. But white dwarfs are so dim that you can't see a single one without using a telescope.

Astronomers discovered the first white dwarf a century ago. At that time, they thought all dim stars were red. But in 1910 two astronomers checked their data and realized that a dim star named Omicron[2] Eridani B was *white*, not red.

"I was flabbergasted," said Henry Norris Russell, the *R* in the H-R diagram. "I was really baffled trying to make out what it meant."

But Harvard astronomer Edward Pickering smiled. "I wouldn't worry," he told him. "It's just these things which we can't explain that lead to advances in our knowledge."

You can see the nearest white dwarf in the photograph above. The big star is Sirius A—the brightest star in the night. The tiny point to its lower left is Sirius B, a white dwarf. It's 8.6 light-years from Earth.

Astronomers first saw Sirius B in the 1800s. But they didn't know it was a white dwarf until the 1900s.

For many years, Prussian astronomer Friedrich Wilhelm Bessel looked at Sirius.

He measured its position. The star should move through space in a straight line. But in 1844, Bessel reported that the star moved first one way, then the other way.

Why? What could make the star wobble? Bessel thought. Suppose a dark star went around Sirius. Then the dark star's gravity would tug Sirius one way and then another, making Sirius wobble.

Bessel looked for this dark star. So did other astronomers. But no one found it—until 1862, when American astronomer Alvan Graham Clark used a large telescope. He saw a tiny point of light beside Sirius. It was Sirius B!

But no one saw Sirius B well. It was too faint and too close to a bright star. So people thought it was just a red dwarf.

Then, in 1915, American astronomer Walter Adams used a larger telescope. He observed the star's spectrum. The spectrum showed the star was too hot to be a red dwarf. Instead, it was a *white* dwarf.

White dwarfs are tiny. Take Sirius B. Its surface temperature is 25,000 Kelvin—about 4 times the Sun's surface temperature. So, according to the Stefan-Boltzmann law, every square inch of Sirius B's surface emits 4 x 4 x 4 x 4 = 256 times more light than every square inch of the Sun's surface. So if Sirius B were as big as the Sun, it would emit 256 times more light.

But Sirius B is faint. It emits much *less* light than the Sun.

How can this be? How can the star be hot yet faint?

There's only one way. Sirius B must be small—much smaller than the Sun. In fact, Sirius B is even a bit smaller than the Earth. Yet it has the same mass as the Sun. So Sirius B is very dense. It is 200,000 times denser than lead. The star has lots of mass packed into a small space.

Because a white dwarf is so dense, the gravity at its surface is immense. If you weigh a hundred pounds on Earth, you'd weigh thousands of *tons* on a white dwarf. If you dropped a rock from a height of four feet, gravity would pull the rock down so fast that it would crash into the white dwarf at 4,000 miles per hour.

White dwarfs have so much gravity that photons—particles of light—lose energy as they climb away from the star. Astronomers see the spectrum of the star shift to the red, the weakest color. We call this the gravitational redshift.

A white dwarf was once a main-sequence star that was born with less than eight times the Sun's mass. Then it expanded into a red giant. The red giant cast its outer layers into space and created a planetary nebula, exposing the red giant's small, dense, hot core. That hot core is the white dwarf.

Unlike the Sun, a white dwarf does not create new energy through nuclear reactions. Instead, it shines from its leftover heat.

Why doesn't the white dwarf collapse? The star's electrons push against each other, holding the star up against its own gravity. This outward force does *not* arise because electrons have the same charge. Instead, it happens whenever a star is so dense that it crams its electrons close together. Two electrons don't "like" to be in exactly the same place, just as two people in a movie theater don't like to sit in the same seat.

A white dwarf obeys a strange rule: the more mass it has, the *smaller* it is—because gravity compresses it more. For example, Sirius B has more mass than Procyon B, another white dwarf. But Sirius B is slightly smaller than Earth, whereas Procyon B is somewhat larger than Earth.

Over time, as a white dwarf sends light into space, it cools and therefore fades. Slowly it turns from blue to white to yellow to orange to red. So despite their name, all white dwarfs aren't white. In fact, on the H-R diagram (pages 18–19), you can see white dwarfs of various colors in a line below the main sequence. There you see Sirius B (blue), Procyon B (white), and Van Maanen's Star (yellow-white).

The hottest, brightest, and bluest white dwarfs—such as Sirius B—are the youngest. They became white dwarfs recently. In 2005, astronomers estimated that Sirius B was a red giant just 124 million years ago. That's when dinosaurs still ruled the Earth.

The astronomers also estimated that Sirius B once had five solar masses. Today it has only one solar mass. So it lost four solar masses when it was a red giant.

The Sun is a future white dwarf. Astronomers calculate that it will become a white dwarf in 7.8 billion years. At that time, the Sun will be faint, and the Earth will become frigid.

Supergiants and Supernovae

One supernova can outshine an entire galaxy.

Supergiants are the most luminous stars in the Galaxy. Some red supergiants are so big that if one were at the center of our solar system, it would swallow all the planets out to Jupiter. Someday these stars will explode as supernovae.

The two most famous red supergiants are Antares, in the constellation Scorpius, and Betelgeuse, in the constellation Orion. These are the brightest red supergiants in our sky. They're in opposite directions from us, just as New York and Denver are in opposite directions from Chicago.

You can find both stars at the upper right (bright and red) of the H-R diagram (pages 18–19). Each star emits about 10,000 times more visible light than the Sun. That means each emits as much light in a minute as the Sun emits in a week.

Both stars are cooler than the Sun. That's why they're red.

But the surface of a cool star emits little light per square inch. Antares and Betelgeuse have a bit more than half the surface temperature of the Sun. According to the Stefan-Boltzmann law, the surface of a star with half the Sun's temperature emits only $1/2 \times 1/2 \times 1/2 \times 1/2 = 1/16$ as much light per square inch as the Sun's surface.

Then how can Antares and Betelgeuse shine so brightly? There's only one way. These stars must be big. Really big. Really, really big—so big that astronomers call them supergiants.

To confirm the huge sizes, astronomers have measured some of these stars' diameters. For example, every now and then, the Moon passes over Antares. Astronomers watch the star's light fade. The bigger the star, the longer the Moon takes to cut off all the star's light. Astronomers find that Antares is 800 times the diameter of the Sun.

Not all supergiants are red. Some, such as Rigel in Orion, are blue. Others, like Deneb in Cygnus, are white. And still others are yellow or orange. They all emit much more light than the Sun.

Supergiants are massive stars. They were born with more than eight times the mass of the Sun.

When such a star is young, it is hot, bright, and blue. It is a main-sequence star of spectral type O or at the hot end of spectral type B. It makes energy the same way the Sun does: its center changes hydrogen into helium.

Huge amounts of both visible and invisible light push outward from the star's center. This is good, because the force of gravity pulls inward and tries to make the star collapse. The gravity is so great that gas pressure alone can't resist it. So the outflowing light helps hold the star up against its own gravity.

The constellation Orion has the red supergiant Betelgeuse (upper left) and the blue supergiant Rigel (lower right).

Orion

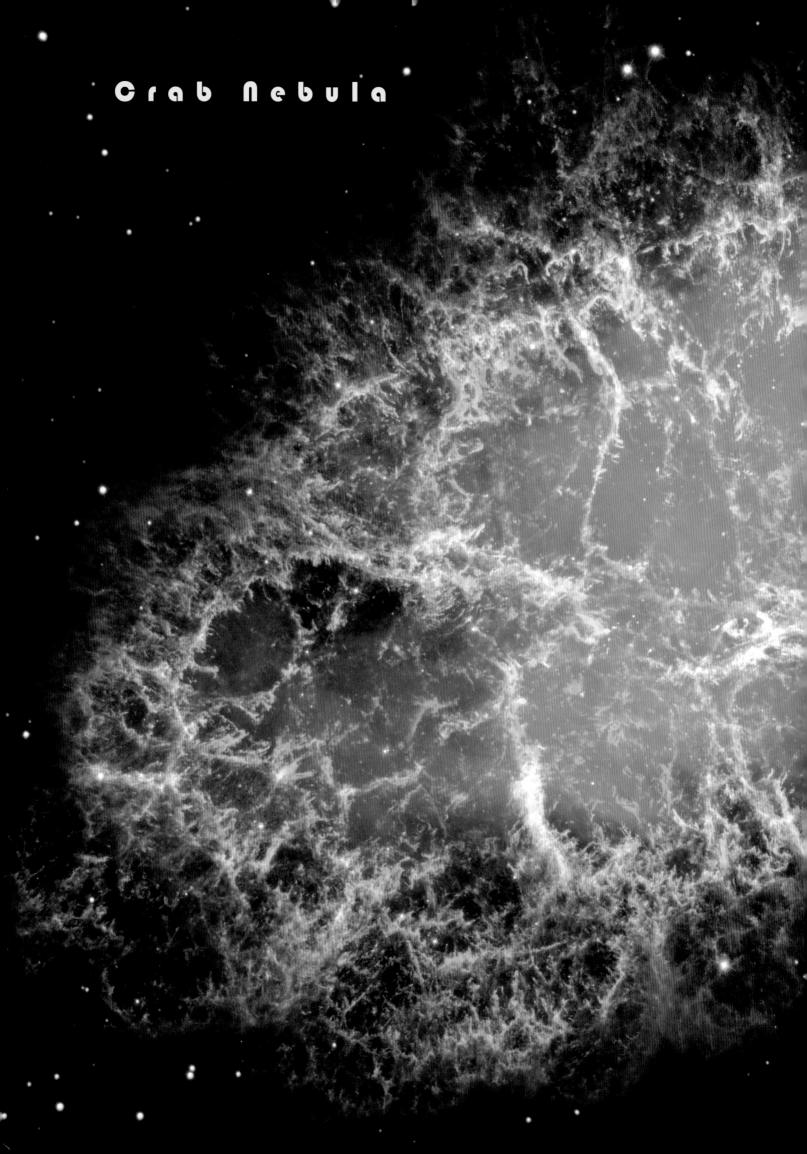

However, the star shines so brightly that it uses up its fuel within millions of years—much less time than the *billions* of years the Sun will take to use up its fuel. First the star's center runs out of hydrogen. In most cases, the star then expands and cools, turning white, then yellow, then orange, then red—until it is a red supergiant like Antares and Betelgeuse.

The red supergiant makes energy by changing helium and other elements into heavier elements. But these nuclear reactions do not release as much energy as hydrogen did. Within a few million years, the star has no fuel left. The star's core becomes iron, and iron is deadly: if two iron nuclei were to fuse together, they wouldn't release any energy. They'd steal it instead.

Now the star is in big trouble. The star can't make energy to hold itself up, but gravity is still trying to pull the star inward. Gravity caused the star's birth; now gravity will cause its death. In less than a second, the star's center collapses. Meanwhile, for reasons astronomers don't fully understand, the star's outer layers shoot into space at millions of miles per hour. The star has exploded!

The explosion is a supernova. The last time people saw a supernova in our Galaxy was 1604. That was before astronomers were using telescopes. However, every year astronomers see supernovae in other galaxies. (*Supernovae* is the plural of *supernova*, just as *nebulae* is the plural of *nebula*.) At its peak, one supernova outshines all the other stars in a small galaxy put together.

In the year 1054, people saw a star explode in the constellation Taurus. It was so bright it cast shadows at night. People saw it during the day. Centuries later, astronomers used telescopes to discover a fuzzy patch of light in Taurus. This is the Crab Nebula. It is the most famous supernova remnant—the remains of the exploded star. It's 6,500 light-years away.

The Veil Nebula in Cygnus is an older supernova remnant. The star that created it exploded about 10,000 years ago. Like a snowplow, the blast swept material through space. Unlike the Crab Nebula, the Veil Nebula is made mostly of this swept-up material, rather than debris from the exploded star. In this way, massive stars return material to the Galaxy that gave them birth.

In this false-color view of the Cassiopeia A supernova remnant, red represents infrared radiation, which comes from room-temperature dust; yellow represents visible light, which comes from gas that is twice as hot as the Sun's surface; and green and blue represent x-rays, which come from extremely hot gas whose temperature is up to 10 million Kelvin. The turquoise-blue dot near the center is the neutron star that was born in the supernova.

Cassiopeia A

As this supernova remnant moves from left to right, it destroys dust grains in space. Dust blocks starlight. So you see more stars per square inch on the left side of this photograph, where the dust has been destroyed, than on the right side.

Veil Nebula

Cepheids

They reveal distances to other galaxies.

The most famous supergiants are red, like Betelgeuse, or blue, like Rigel. But yellow supergiants called Cepheids are the most useful. Cepheids (SEH-fee-idz) reveal distances to other galaxies.

A Cepheid is a type of variable star. It looks brighter on some nights than on other nights.

Thousands of years ago, people thought stars couldn't vary. Greek philosopher Aristotle had said the heavens were perfect and changeless.

They aren't, of course, and in the 1780s came the discovery of the first two Cepheids. At that time astronomers knew just a few variable stars. One was the red star Mira. But an English gentleman named Edward Pigott thought there might be many more.

Night after night, Pigott compared one star with another. He hoped to find a star that brightened or faded.

During his search, Pigott guided John Goodricke, a young man living near him. Goodricke, who was eleven years younger than Pigott, could not hear or speak.

At that time, many people thought a deaf-mute like Goodricke was stupid. They said language separated people from animals. So those who could not speak must be as stupid as animals; they should not even go to school.

Fortunately, Goodricke's parents thought otherwise. They made sure Goodricke went to school. And in 1781, at the age of seventeen, Goodricke began observing the sky with Pigott.

On the night of September 10, 1784, Pigott discovered that the light of the star Eta Aquilae was varying. Pigott saw the star brighten fast, then fade slowly, once a week.

A month later, on October 20, Goodricke discovered that one of the stars in the northern constellation Cepheus was variable. But which star? Three nights later, on October 23, Goodricke became "*almost* convinced" that the varying star was Delta Cephei. Like Eta Aquilae, Delta Cephei brightened fast, then faded slowly.

From England, Delta Cephei never sets. So Goodricke could observe it every clear night. He soon found that its brightness varied every five days.

Unfortunately, while observing Delta Cephei, Goodricke became ill. He died of pneumonia in 1786, only twenty-one years old. "I had the misfortune to lose the best of friends . . . which took away the pleasure I ever had in astronomical pursuits," Pigott wrote. Pigott himself died many years later, in 1825, unhappy that he had received little credit for his discoveries.

Yet the two stars Pigott and Goodricke identified—now called Cepheids, after Delta Cephei—would prove crucial to modern astronomy.

Most Cepheids vary with precision. For example, Eta Aquilae, the first Cepheid discovered, goes from bright to dim and back to bright again every 7.1767 days. Delta Cephei does the same every 5.3663 days.

Even Polaris, the North Star, is a Cepheid. However, the North Star's light changes so little that you'd need a telescope to see the variation. Still, Polaris is the closest and brightest Cepheid. It is 430 light-years from Earth.

In 1907, American astronomer Henrietta Leavitt, who was nearly as deaf as Goodricke, made a big discovery. Leavitt measured the periods of sixteen Cepheid variables in a nearby galaxy. "It is worthy of notice," she wrote, "that . . . the brighter variables have the longer periods."

All the stars in this galaxy were nearly the same distance from Earth. That meant the bright Cepheids not only looked brighter but also emitted more light into space than the faint Cepheids. Thus, Cepheids with long periods must be more luminous than Cepheids with short periods. As we'll see, this trait means Cepheids reveal distances.

Today astronomers know Cepheids are pulsating yellow supergiants. As the stars expand and shrink, they brighten and fade. The bigger and brighter they are, the more slowly they pulsate.

Cepheids are like musical instruments in an orchestra. A big musical instrument, such as a tuba, emits a low pitch, a sound that vibrates slowly. A small musical instrument, such as a trumpet, emits a higher pitch, a sound that vibrates fast. In the same way, a big and bright Cepheid pulsates slowly—once every month or two—whereas a smaller and fainter Cepheid pulsates fast, once every day or two.

Even if you can't see a musical instrument, you can often tell its size just by hearing its pitch. Likewise, even though astronomers don't at first know how far a Cepheid is, they can tell how much light it emits into space by measuring its period. Comparing how much light the star emits with how bright the star looks reveals its distance.

Furthermore, Cepheids are so luminous they can be seen in galaxies far beyond the Milky Way. As a result, astronomers can measure the distance of a galaxy by figuring out the distance of the Cepheids in it.

In this way, astronomers have measured the distances to galaxies tens of millions of light-years away. For example, the Andromeda Galaxy, the nearest giant galaxy to our own, is 2.5 million light-years away. And spiral galaxy NGC 3370, shown here, is 90 million light-years from Earth—three dozen times farther than Andromeda.

Neutron Stars and Pulsars

A pulsar is like a fast-spinning lighthouse.

When a massive star explodes as a supernova, it shoots its outer layers into space. But the star's center collapses—and may become a tiny star made of neutrons. Astronomers see some neutron stars as pulsars.

Physicists discovered neutrons in 1932. Neutrons are particles that are in most atoms on Earth. A neutron is like a proton, except it has no electric charge. In contrast, a proton has positive charge, and an electron has negative charge.

Could there be stars made only of neutrons?

Two astronomers in California—Walter Baade and Fritz Zwicky—were thinking about stars that explode in other galaxies. They were thinking about how bright those supernovae were. The collapse of a star might release enough gravitational energy to make a supernova shine brightly.

In 1934, Baade and Zwicky wrote, "With all reserve we advance the view that a super-nova represents the transition of an ordinary star into a *neutron star*, consisting mainly of neutrons. Such a star may possess a very small radius and an extremely high density."

They were right. But for a long time, no one knew. Then, in 1967, Jocelyn Bell, an astronomy student in England, found something strange in the sky.

When Bell was a child, she liked astronomy. But she almost gave it up, because she didn't want to work at night.

However, astronomers who study radio waves don't have to work at night. Radio telescopes detect radio waves during the day, too.

One day Bell found something in the constellation Vulpecula. It sent out a pulse of radio waves every $1^{1/3}$ seconds. So astronomers called it a pulsar.

It was the first pulsar ever seen. But what was it? Was it just a problem with the radio telescope?

No, another radio telescope detected the same thing.

Was the pulsar a message from another planet?

No. Such a planet should move around its star. This movement would change the wavelength of the radio waves astronomers measured. But astronomers saw no change in the wavelength.

Astronomers soon figured out what the pulsar was: a fast-spinning neutron star. The pulsar spins every $1^{1/3}$ seconds—much faster than the Sun, which spins once a month.

A pulsar is like a lighthouse. A lighthouse emits light in a beam. Every time its beacon spins, sailors see a flash of light.

Likewise, a pulsar emits radio waves in a beam. Every time it spins, radio astronomers see a flash, or pulse, of radio waves.

A neutron star is born when a massive star dies. What had been one of the biggest stars in the Galaxy—a red supergiant—becomes one of

This false-color image shows the center of the Crab Nebula. Red represents visible light and blue represents x-rays. The arrow points to the Crab pulsar.

Crab Pulsar

the smallest. When the massive star explodes as a supernova, the star's center collapses. Positively charged protons smash into negatively charged electrons, creating neutral neutrons. Although neutrons have no electric charge, they do push against one another when crammed together. This pressure holds the neutron star up against the force of its own gravity.

Furthermore, the collapse makes the star spin fast. In the same way, spinning ice skaters spin faster when they draw in their arms.

A typical neutron star has $1^1/_2$ times the mass of the Sun. But it's only ten miles across. That's smaller than some cities on Earth.

So a neutron star is very dense. A spoonful weighs more than a billion tons. If you drop a pebble from a height of four feet above a neutron star, the gravity is so strong it would cause the pebble to crash into the surface at 5 million miles per hour with the energy of thousands of sticks of dynamite.

Astronomers see some pulsars in supernova remnants. The most famous pulsar is in the Crab Nebula, the supernova remnant that formed when our ancestors saw a star explode in 1054. So that pulsar is about a thousand years old. It spins 30 times a second.

As a pulsar gets older, the supernova remnant expands and thins out, until it vanishes from view. Therefore, most pulsars are not in supernova remnants. And as time goes on, the pulsar spins more slowly, until—about 10 million years after its birth—it stops sending out radio waves.

Since Bell's discovery of the first pulsar, astronomers have found thousands of other pulsars. But there must be millions of dead pulsars, too: tiny but dense neutron stars that roam the Galaxy.

Neutrons hold a neutron star up against the force of its own gravity. But what if the star has too much mass? Then the star collapses completely—and becomes a black hole.

Black Holes

Cygnus X-1

Nothing can escape—
not even light.

Artist's interpretation

A black hole is a star that doesn't shine. It has so much gravity that nothing, not even light—the fastest thing in the universe—can escape.

A black hole is the remains of a massive star that has collapsed. All massive stars—those born with more than eight times the mass of the Sun—explode as supernovae. Many of these stars collapse into neutron stars, in which tightly packed neutrons push against one another and hold the star up against the inward-pulling force of gravity. A typical neutron star has just $1^{1}/_{2}$ solar masses, because a massive star casts lots of its mass into space both during its life and during its death, in the supernova explosion.

But if the collapsed star has more than three times the mass of the Sun, the gravity is just too strong. Not even neutrons can hold the star up. So the star collapses further—into a black hole.

Why is it called a black hole? Well, because light can't escape it, it doesn't shine. So it's black. And because once you fall into it, you can't get out, it's a hole. Put the two together and you have *black hole*.

All the matter in the collapsed star is at the center of the black hole. But the gravity of that material reaches much farther out. Surrounding a black hole is the "event horizon," the point of no return. Anything—or anyone—that falls through the event horizon won't be able to get out of the black hole.

There must be millions of black holes scattered throughout the Milky Way. After all, lots of massive stars have died and turned into black holes during the long life of the Galaxy.

Trouble is, black holes are hard to find. Unlike normal stars, they give off no light. But by being clever, astronomers have discovered many black holes in space.

In 1971, astronomers found the first black hole. They launched a spacecraft that went around the Earth. This satellite searched for x-rays. X-rays travel at the speed of light but have much more energy. However, the Earth's air blocks x-rays from space, which is why astronomers had to launch a satellite to detect them.

The satellite observed x-rays coming from the constellation Cygnus. Earlier rocket flights had also seen these x-rays. It was the first x-ray source found in Cygnus. So astronomers called it Cygnus X-1.

But what was making the x-rays? Astronomers used radio telescopes on Earth to look at Cygnus. Radio telescopes pinpoint positions better than x-ray satellites do. The radio telescopes found that Cygnus X-1 also emitted radio waves, which were coming from near a blue star in Cygnus.

Astronomers then studied the blue star by using telescopes that detect visible light. They discovered that the blue star was speeding around something. But no one could see the something. It had to be a dark star—possibly a black hole.

The blue star was moving fast around the dark star. In order to force the blue star around it so fast, the dark star must have a lot of gravity and therefore a lot of mass. Astronomers calculated: the dark star had so much mass that it was indeed a black hole.

What about the x-rays? Those are what tipped astronomers off to the black hole. Because x-rays have lots of energy, they are coming from something much hotter than the Sun. Suppose gas flows from the blue star toward the black hole. This gas falls fast toward the black hole and spins around it. The gas rubs against other gas, so it gets very hot—and emits x-rays before it falls into the black hole.

Imagine you tried to go into a black hole. What would happen? What would you see?

Well, it's not a good idea. First, even when you're outside the event horizon, the black hole's gravity is so great that you would get torn apart by tides.

Tides happen when gravity pulls more strongly on one side of an object than on the other side. For example, the Moon's gravity pulls more strongly on the side of Earth that faces it than on the other side of Earth. The ocean on the side of Earth facing the Moon therefore gets pulled up above the ground.

A black hole has a lot more mass than the Moon. Plus, a black hole such as Cygnus X-1 is much smaller than the Moon, so you can get much closer to the black hole's center than you can to the Moon's center. So the tides near a black hole are much stronger. If you dive through the black hole feetfirst, the black hole's tides would tear your feet from your legs, your legs from your torso, and your torso from your head. That's because the gravity pulling on your feet is much stronger than the gravity pulling on your head. So you see why going into a black hole isn't smart!

Still, let's pretend you survive the black hole's tides. You dive through the event horizon. Most of the space inside the event horizon is probably empty. Nearly all the matter in the black hole is at the very center.

Who knows what you would discover? Maybe you will find a path to another part of the universe or even to another universe altogether.

Suppose you try to tell the rest of us what you find. You send us radio signals. Unfortunately, your radio signals travel at the speed of light. Since light can't get out of the black hole, neither can your radio signals. The gravity is too strong. So the black hole remains a mystery to everyone—except you, the brave explorer.

Double Stars

The brightest star at night is double.

Albireo

Our Sun is a single star. No star goes around the Sun.

But many stars are double. Some stars are triple. And a few stars are quadruple (four stars), quintuple (five stars), and even sextuple (six stars).

The brightest star of the night is double. Sirius is really two stars: a bright main-sequence star, named Sirius A, and a dim white dwarf, Sirius B. You can see their picture on page 32.

Together, the two stars are called a star system. Sirius A and B go around each other every 50 years.

The closest star to the Sun is also a multiple star system. Alpha Centauri is triple. Its three stars go around one another.

The brightest star in the Alpha Centauri system is called Alpha Centauri A. It is a yellow G-type main-sequence star, like the Sun. Near this star is Alpha Centauri B, an orange K-type main-sequence star that is somewhat fainter than the Sun. On average the two stars are about 2 billion miles apart, a bit farther than Uranus is from the Sun. Alpha Centauri A and B go around each other every 80 years. They are 4.37 light-years from the Sun.

Far from these two stars is a dim red dwarf, Alpha Centauri C. It is 13,000 times farther from Alpha Centauri A and B than the Earth is from the Sun. Because it is so far away from the two main stars, it probably takes a million years to go around them.

Alpha Centauri C is slightly closer to us than Alpha Centauri A and B are. It is 4.24 light-years from Earth. So astronomers often call this small star Proxima Centauri, because *proximity* means "closeness."

You have to be careful if someone asks you what the closest star is. The closest *star system* to the Sun is Alpha Centauri. The closest individual *star* to the Sun is Proxima Centauri, which is part of the Alpha Centauri star system. And, of course, the closest star to the *Earth* is the Sun!

Stars usually go around each other on elliptical, or oval-shaped, paths. For example, at their closest, Alpha Centauri A and B are 1 billion miles apart. That's a bit farther than Saturn is from the Sun. But at their farthest, the two stars are more than 3 billion miles apart. That's a bit farther than Neptune is from the Sun.

Two stars orbit each other by going around their center of mass. This is an imaginary point in space between the two stars. It is always closer to the more massive star. To picture it, imagine placing both stars on a giant seesaw. The center of mass is where you would put the fulcrum in order to balance the seesaw.

For example, Sirius A is twice as massive as Sirius B. So the center of mass lies twice as close to Sirius A as it does to Sirius B. Each star goes around the center of mass—but because Sirius A is more massive, Sirius A moves less than Sirius B.

Of the bright stars, Castor in Gemini is a most amazing multiple star. It has six stars! Four stars are white, and the other two are red dwarfs.

Double stars are nice to look at through telescopes. For example, Albireo, in Cygnus, is a beautiful double star made of an orange and a blue star.

Algeiba in Leo is also beautiful. It's made of two giant stars—one yellow, the other orange—that go around each other.

Double stars are especially important because they help astronomers measure stellar masses. That's because each star feels the other's gravity, whose strength depends on how much mass the star has. The more mass the stars have, the faster they move around each other. So by measuring a double star's orbit, astronomers can figure out the masses.

Some double stars eclipse each other. The most famous example is Algol in Perseus. Algol's two brightest stars are blue and orange. The blue star is brighter than the orange star, but the orange star is bigger. They orbit each other every 2 days and 21 hours. When the big orange star passes in front of the smaller blue star, the whole system dims. You can even see these eclipses without using binoculars or a telescope. Astronomers call a system like Algol an eclipsing binary.

Double stars can do things single stars don't. For example, two stars may be so close together that one star spills some of its mass onto the other. In the Algol system, the big orange star is only a few million miles from the blue star—much closer than Mercury is to the Sun. As a result, the orange star spills gas onto the blue star.

This flow of gas explains a puzzle. The orange star has much less mass than the blue star. So the orange star should age more slowly. But in fact, the orange star is already becoming a giant—while the blue star is still on the main sequence.

How can this be? Astronomers think the orange star was born with more mass than the blue star. But when the orange star left the main sequence and expanded, it dumped so much material onto the other star that its partner is now the more massive star.

If one star in a double star system is a white dwarf, material that falls onto the white dwarf can explode. Then we see a nova in the sky. *Nova* means "new," because we see what looks like a new star. But it's not really a new star. It's just a flare-up in a double star system that has a white dwarf. And neither star gets destroyed.

But one kind of double-star explosion *does* destroy a star. These explosions are so luminous that astronomers can see them from a distance of billions of light-years: type Ia supernovae.

When Little Stars Explode

They can be seen across the universe.

A supernova explosion can mark the death of a massive star—one born with more than eight times the mass of the Sun. Such a star usually swells into a red supergiant and then explodes, leaving behind a neutron star or black hole.

But small stars can explode, too. These stars are smaller than the Sun. They are smaller than the Earth. But when they explode, they shine very brightly.

Astronomers call them type Ia supernovae. For a long time, these supernovae were mysterious. Astronomers used telescopes to observe supernovae in other galaxies and saw several strange things. First, spectra showed that type Ia supernovae have no hydrogen—even though hydrogen is the most common element in the universe.

Plus, astronomers saw type Ia supernovae in all kinds of galaxies. In contrast, other types of supernovae explode only in galaxies with young stars. That makes sense. Massive stars burn their fuel fast. So they die young, and their supernovae appear only in galaxies with young stars. Because type Ia supernovae occur in all kinds of galaxies, including galaxies with no young stars, the type Ia explosions were not coming from short-lived massive stars.

Furthermore, astronomers noticed that most type Ia supernovae were very similar. The supernovae brightened, then faded, with nearly the same pattern. In contrast, other types of supernovae differ from one explosion to the next.

Astronomers thought about these puzzles. They came up with an idea. They said type Ia supernovae are exploding white dwarf stars.

This idea explains a lot. Most white dwarfs are made of carbon and oxygen, with little or no hydrogen. So when a white dwarf explodes, astronomers see no hydrogen in its spectrum.

In addition, many white dwarfs are old. After all, it takes time for a main-sequence star to become a red giant and then a white dwarf. So type Ia supernovae can happen even in galaxies having only old stars.

Why does a white dwarf explode? Most white dwarfs don't. For example, Sirius B, the closest white dwarf to Earth, won't ever explode.

However, if a white dwarf is close to another star, and if that other star dumps material onto it, there might be trouble.

Here's why. White dwarfs are dead stars. They make no energy. Instead, they shine from the leftover heat of days past.

What holds white dwarf stars up against the force of their own gravity? The answer: pressure between the white dwarf's electrons. This is the same sort of pressure that neutrons exert to hold up a neutron star.

But this electron pressure can exert only so much force. If a white dwarf's mass exceeds more than 1.4 times the mass of the Sun, its gravity gets too strong, and the star explodes!

Most white dwarfs are safe. They're well below the danger point. The typical white dwarf has just 60 percent of the Sun's mass. But some white dwarfs are more massive. If another star dumps material onto a massive white dwarf, then the white dwarf may face disaster.

When the star's mass reaches 1.4 solar masses, carbon nuclei in the white dwarf start to join other carbon nuclei. This nuclear reaction makes energy. The energy makes more carbon nuclei join together. Then there's even more energy—which makes even more carbon nuclei fuse together, and so on.

As a result, a runaway nuclear explosion rips the white dwarf apart. Nothing is left—no neutron star, no black hole—just the shattered remains of the exploded star that rush out in all directions.

Now you know why all type Ia supernovae look nearly the same: because they all come from white dwarfs with 1.4 solar masses.

Type Ia supernovae are so bright that astronomers see them in galaxies billions of light-years away. In fact, because type Ia supernovae are so similar to one another, astronomers use these explosions to measure the distances of faraway galaxies: the fainter the supernova looks, the farther its galaxy is from Earth.

When a white dwarf goes supernova, it makes lots of iron. In fact, most of the iron in your blood came from exploding white dwarfs.

A type Ia supernova (lower left) explodes in a galaxy.

Star Clusters

Some stars travel with many others.

Pleiades

Most stars sail through space alone or with just a few other stars. Some stars, though, belong to beautiful clusters.

The best-known star cluster, shown here, is the Pleiades. It is 435 light-years away in the constellation Taurus. If you look at the Pleiades through binoculars, you will see a stunning sight.

Taurus has another famous star cluster: the Hyades, 150 light-years from Earth. The Hyades is the closest star cluster to Earth.

Both the Pleiades and the Hyades are open star clusters. A typical open cluster has hundreds of stars. These stars are spread out from one another, so the gravity of one star on another star is weak. But this same gravity holds the cluster together.

Over time, as giant clouds of molecular gas pass by the cluster, their gravity tears stars away—until the cluster breaks apart. Thus, open star clusters don't last as long as the Galaxy.

As a result, most open star clusters are young. The Milky Way has about 1,200 known open star clusters. All but a few dozen are younger than a billion years. Very few are older than the 4.6-billion-year-old Sun.

Long ago, the Sun may have been in a star cluster like the Pleiades. But the cluster got torn apart. Now the Sun goes through the Galaxy alone.

After a star cluster disperses, its stars still move through space in the same direction, like parachutists jumping out of the same airplane. In 1869, British astronomer Richard Proctor

An open cluster

discovered that five of the Big Dipper's stars moved through space in the same direction.

Today astronomers know stars all over the sky that share this motion. These stars belong to the Ursa Major moving group, so named because the Big Dipper is in the constellation Ursa Major. These stars were probably born in an open star cluster that broke apart.

Open clusters are the more common type of star cluster. But some star clusters are a different type. They are called globular clusters, because they have the round shape of a globe. A typical globular cluster has hundreds of thousands of stars, packed tightly together.

The Galaxy's most luminous globular cluster is Omega Centauri. Altogether its stars emit a million times more light than the Sun.

Astronomers have found only about 150 globular clusters in our Galaxy. One of the nearest globular clusters is M4 in Scorpius. M4 is 7,000 light-years away—much farther than the nearest open star clusters.

Because globular clusters are crammed with stars, the stars' gravity can hold the cluster together for many billions of years. In fact, all globular clusters in the Milky Way are old. Many are as ancient as the Galaxy—13 or 14 billion years old.

Astronomers can tell how old a cluster is by looking at its brightest main-sequence stars. For example, the Pleiades cluster is young. Its brightest stars are blue B-type stars. Such stars don't live long, so the cluster must have formed recently. Astronomers estimate the Pleiades is only about 100 million years old.

The Hyades cluster is older. All of its blue B-type stars have died. Its brightest main-sequence stars are white, of spectral type A. Astronomers think the Hyades is 600 million years old. That's still much younger than the Sun.

But in a globular cluster, the brightest main-sequence stars are yellow stars like the Sun. A globular cluster is so old that all of its O, B, and A main-sequence stars have died.

The brightest stars in a globular cluster are orange and red giants. These stars were once main-sequence stars similar to the Sun.

By studying star clusters, astronomers can see how the Galaxy has changed over time. The oldest star clusters—the globulars—have few of the heavy elements, such as oxygen and iron, that life needs. Thus, 13 billion years ago, when many of the globular clusters formed, the Galaxy itself had few of these heavy elements.

In contrast, the young open clusters usually have lots of heavy elements. Because these clusters arose only recently, the Galaxy must now have more heavy elements than it did long ago.

Indeed, the Milky Way now has so many heavy elements that life can exist. We therefore owe our existence to the creators of those heavy elements: the stars themselves.

A globular cluster

M 8 0

Origin of the Elements

We are made mostly of elements created in the stars.

The air you breathe, the water you drink, the food you eat all have atoms made by the stars. Without stars, we wouldn't be here. Neither would the Earth.

Billions of years ago, long before the Sun and the Earth were born, the only elements in space were the three lightest: hydrogen, helium, and a little bit of lithium. These three elements arose in the big bang that gave birth to our universe 14 billion years ago.

But those three elements can't make life. Life needs heavier elements—like carbon, the sixth lightest element, which is the basis of all life on Earth; nitrogen, the seventh lightest element, which is in protein; and oxygen, the eighth lightest element, which we breathe.

All these heavy elements came from the stars.

In two ways, astronomers figure out which stars made which elements. First, astronomers calculate how different types of stars make different elements during their lives and how the stars cast these elements into space when they die.

Second, astronomers observe the mixture of elements on a star's surface, measuring the percentage of hydrogen, the percentage of helium, the percentage of carbon, and so on. This surface mixture is usually what the star was born with. Thus, an old star, which formed long ago, reveals the mixture of elements the Galaxy had long ago. And a young star, which formed recently, shows what the Galaxy has today. By comparing the surfaces of old and young stars, astronomers see which elements entered the Galaxy first.

For example, they find that oxygen entered the Milky Way almost as soon as our Galaxy was born, so the stars that make oxygen must die soon after they are born. In contrast, iron entered the Galaxy slowly, so the stars that make iron must live longer.

In these ways, astronomers can say how most elements arose. For example, the nitrogen that makes up most of Earth's air came from stars that did not explode. These stars were born with less than eight times the mass of the Sun. During their lives, these stars made nitrogen. When they died, they cast their outer layers into space, creating planetary nebulae like the beautiful Ring Nebula. As the planetary nebulae expanded, they carried the new nitrogen into space.

In contrast, most oxygen was made by massive stars, those born with more than eight times the mass of the Sun. During their lives, these stars combine four helium nuclei to make one oxygen nucleus. Then, when the stars explode as supernovae, they cast the oxygen into space. You are breathing this oxygen right now.

Carbon arose in both ways.

Iron has yet another origin. Most iron on Earth—and in your blood—was forged when white dwarfs exploded as type Ia supernovae. However, a lot of iron also came about when massive stars exploded as supernovae.

Elements much heavier than iron—such as gold and platinum—are rare. For every gold atom in space, there are 70 million oxygen atoms and 120 billion hydrogen atoms. That's because stars don't make much gold. Normally, as stars change light elements into heavier ones, they release energy. For example, the Sun shines because it's turning hydrogen into helium.

But this process goes only so far. It doesn't go beyond iron. Changing iron into a heavier element *absorbs* energy from the star.

Still, stars do create elements heavier than iron. They do this in two main ways.

First, when a massive star explodes, neutrons smash into some of the iron nuclei outside the star's core, changing them into heavier elements. Scientists call this the r-process (the *r* stands for *rapid*), because the neutrons hit the iron nuclei rapidly. Neutrons join the nuclei because they have no electric charge, so the nuclei—which contain protons and thus have positive charge—don't push them away.

The r-process created many of the elements heavier than iron, including nearly all the gold, platinum, iodine, and uranium in the universe. So if you see some gold jewelry, think of the supernova explosions that forged the element.

Second, when a red giant or supergiant makes energy, it can release neutrons that slowly change some of its iron nuclei into heavier elements. This is the s-process—the *s* stands for *slow*. The s-process made most of the lead in the universe.

After stars make their elements, they cast them into space. The elements sprinkle the nebulae where stars can form.

And 4.6 billion years ago, one of those nebulae gave birth to the Sun and its planets. The Sun is mostly hydrogen and helium, elements made mostly in the big bang. But the Earth is mostly oxygen, silicon, and iron, elements made by the stars.

And you too are made mostly of elements made in the stars: the calcium in your bones, the nitrogen in your protein, the oxygen you breathe, the iron and copper in your blood.

So the stars do more than shine at night. They also make life possible—life that can look at the stars, admire their beauty, and figure out how they created the elements on Earth.

The Elements

Element	Symbol	Atomic Number	Made Mostly by
Hydrogen	H	1	Big bang
Helium	He	2	Big bang
Carbon	C	6	Helium burning in red giants
			Helium burning in massive stars
Nitrogen	N	7	Hydrogen burning in main-sequence stars and red giants
Oxygen	O	8	Helium burning in massive stars
Neon	Ne	10	Carbon burning in massive stars
Sodium	Na	11	Carbon burning in massive stars
Magnesium	Mg	12	Carbon and neon burning in massive stars
Aluminum	Al	13	Carbon and neon burning in massive stars
Silicon	Si	14	Oxygen burning in massive stars
Phosphorus	P	15	Neon burning in massive stars
Sulfur	S	16	Oxygen burning in massive stars
Potassium	K	19	Oxygen burning in massive star supernovae
Calcium	Ca	20	Oxygen and silicon burning in massive stars
Iron	Fe	26	Type Ia supernovae
			Massive star supernovae
Copper	Cu	29	s-process in massive stars
Silver	Ag	47	r-process in massive star supernovae
Tin	Sn	50	s-process in red giants
			r-process in massive star supernovae
Iodine	I	53	r-process in massive star supernovae
Platinum	Pt	78	r-process in massive star supernovae
Gold	Au	79	r-process in massive star supernovae
Lead	Pb	82	s-process in red giants
Uranium	U	92	r-process in massive star supernovae

*The abundance says how many atomic nuclei of the element existed in the newborn Sun for every million silicon nuclei there. Abundance data are from Katharina Lodders at Washington University: *Astrophysical Journal*, volume 591, page 1220 (2003).

Cast into Space by	Abundance*
Big bang	24,310,000,000
Big bang	2,343,000,000
Planetary nebulae	7,079,000
Massive star supernovae	
Planetary nebulae	1,950,000
Massive star supernovae	14,130,000
Massive star supernovae	2,148,000
Massive star supernovae	57,510
Massive star supernovae	1,020,000
Massive star supernovae	84,100
Massive star supernovae	1,000,000
Massive star supernovae	8,373
Massive star supernovae	444,900
Massive star supernovae	3,697
Massive star supernovae	62,870
Type Ia supernovae	838,000
Massive star supernovae	
Massive star supernovae	527.0
Massive star supernovae	0.4913
Planetary nebulae	3.733
Massive star supernovae	
Massive star supernovae	0.9975
Massive star supernovae	1.357
Massive star supernovae	0.1955
Planetary nebulae	3.234
Massive star supernovae	0.024631

Extrasolar Planets

Like the Sun, many stars have planets.

For a long time, people wondered if stars have planets going around them. After all, the Sun is a star. It has planets. But only in recent years have astronomers discovered planets around other stars.

Planets beyond our solar system are called extrasolar planets. As a prefix, *extra* means "beyond." Just as *extraordinary* means "beyond the ordinary," so *extrasolar* means "beyond the solar system."

Extrasolar planets are hard to detect. That's because planets make no light of their own. They merely reflect the light of their star. Besides Earth, only four of the Sun's planets—Venus, Mars, Jupiter, and Saturn—are easy to see.

The farthest bright planet, Saturn, is less than a billion (1,000,000,000) miles from the Sun. But if Alpha Centauri, the nearest star system, has planets, they're 25 *trillion* (25,000,000,000,000) miles from us. That's over 25,000 times farther than Saturn.

Worse, a star's glare washes out the light of its planets. In our solar system, the biggest planets, Jupiter and Saturn, are about a billionth as bright as the Sun. Imagine how hard Jupiter and Saturn would be to see if you were in another solar system.

In 1991, however, two astronomers made a big discovery. They found planets going around a distant star.

Alex Wolszczan and Dale Frail found the planets in a strange place. These planets did not go around a star like the Sun. Instead, the planets went around a pulsar—a fast-spinning dead star that emits deadly radiation in a beam.

The astronomers did not see the planets. Instead, they watched the pulsar's pulses. Sometimes the pulses came slightly earlier than they should have. Other times, the pulses came slightly later.

Why was this happening?

The astronomers thought.

Perhaps the pulsar had planets—whose gravity tugged the pulsar slightly. When the pulsar moved away from us, its pulses took slightly longer to reach Earth. When the pulsar moved toward us, its pulses took slightly less time to reach Earth.

Astronomers have found three planets around this pulsar. The pulsar is in the constellation Virgo. Its name is PSR B1257+12.

These planets probably have no life. After all, the pulsar emits deadly radiation. But maybe what's deadly to us is life-giving to someone else.

Four years later, in 1995, two other astronomers discovered the first planet going around a star like the Sun.

If the sky is dark, and you know where to look, you can just barely see this star without using binoculars or a telescope. Its name is 51 Pegasi. It's in the constellation Pegasus. It is a yellow G-type main-sequence star.

Swiss astronomers Michel Mayor and Didier Queloz observed this star. They split its light into

Artist's interpretation of a giant planet orbiting a red dwarf star

different colors to see its spectrum. Sometimes the star's spectral features shifted slightly toward longer wavelengths—what astronomers call a redshift, because red is the longest wavelength our eyes can see. Other times the spectral features shifted toward shorter wavelengths—a blueshift. The astronomers found a pattern. The time from redshift to blueshift and back to redshift was 4¼ days.

What was causing 51 Pegasi's redshifts and blueshifts? The astronomers thought. Maybe ... a planet! Suppose this planet goes around the star every 4¼ days. Suppose its gravity tugs the star. When the star moves away from us, we see a redshift in the star's light, because the light waves get stretched slightly. Likewise, when the planet's gravity causes the star to move toward us, we see a blueshift—because the star's light waves get scrunched up, reducing the wavelength slightly.

From the size of the redshifts and blueshifts, the astronomers calculated that the planet was big, like Jupiter. But unlike "our" Jupiter, this Jupiter is very close to its star. In our solar system, the closest planet, Mercury, takes 88 days to go around the Sun. But 51 Pegasi's planet takes just 4¼ days to go around its star. So this planet is much closer to 51 Pegasi than Mercury is to the Sun.

Thus, the planet must be very hot—even hotter than Mercury and Venus. Yet it's a big planet like Jupiter. Astronomers call such planets "hot Jupiters."

Astronomers have now discovered hundreds of planets around other stars. Most of these planets are big, like Jupiter and Saturn. They probably do not have life. But maybe some of their moons do.

Someday, though, astronomers may find a planet the size of Earth going around a star like the Sun. That planet might have air. It might have water. It might even have life!

Life in Space?

Which stars have planets with life?

The Earth is just one planet going around just one star. But our Galaxy has hundreds of billions of stars. Many have planets. Could one of those planets have life? And could some of that life be intelligent?

No one knows—yet.

In our solar system, we know only one world with life: the Earth. Astronomers once hoped Venus had life. Venus has clouds, which suggested rain and water. But when astronomers studied Venus, they found it much too hot and dry.

Mars also gave astronomers hope. Some thought Mars had canals, dug by Martians. But when spacecraft flew past Mars, they found that Mars is a cold, dry desert, probably lifeless.

Europa, an ice-covered moon of Jupiter, might have life. The ice probably covers an ocean of water. Perhaps creatures are swimming there now.

Which stars beyond our solar system might have planets with life? Let's try to answer this question by thinking about the star we know best: the Sun. After all, the Sun has a planet—the Earth—with life. So maybe a star like the Sun would be a good place to look for life.

The Sun is a yellow G-type main-sequence star. Such stars make lots of light, which life needs. But these stars do not burn their fuel as fast as more luminous stars, such as the blue O and B stars or the white A stars. As a result, G-type main-sequence stars live for billions of years. On Earth, that was enough time for life to arise.

Unfortunately, only 1 in 25 stars is a G-type main-sequence star. The two nearest are Alpha Centauri A, 4.37 light-years away, and Tau Ceti, 11.9 light-years away.

The Sun is stable. Its light does not change much. That's good. If a star's light changes a lot, it can freeze and fry an otherwise good planet. Fortunately, most G-type main-sequence stars are stable, like the Sun.

The Sun is 4.6 billion years old. That means it is middle-aged. Some stars are younger than the Sun. Some stars are older than the Sun. The Earth is the same age as the Sun. So intelligent life took 4.6 billion years to arise on Earth.

If life takes more than a billion years to develop, then young stars may not have planets with life. Fortunately, many stars in the Galaxy are middle-aged, like the Sun, or old. Old stars have been around even longer than the Sun and the Earth.

The very oldest stars, however, have a problem. They don't have many heavy elements, such as oxygen and iron. Planets like Earth and living beings like us need heavy elements. But the oldest stars were born before many other stars were able to enrich the Galaxy's nebulae with supernova explosions and planetary nebulae.

Compared with the oldest stars, the Sun has a lot of heavy elements. And Alpha Centauri, the nearest star system, has even more.

Finally, the Sun is a single star. Single stars can have planets. Planets form in a disk of gas and dust around a star. In a double or triple star, one star's gravity may wreck the disk of gas and dust going around another star. Then no planets arise.

In fact, Alpha Centauri has just this problem. The two brightest stars, Alpha Centauri A and B, are yellow G and orange K main-sequence stars. But they are so close together that they may never have formed planets.

Still, *if* planets formed, each star could have planets like Mercury, Venus, Earth, and Mars. Furthermore, astronomers have discovered planets around some double stars.

So who knows? Maybe Alpha Centauri has planets—and possibly life, because the star is slightly older than the Sun.

Will astronomers ever find life in space? Since 1960, they have been searching for radio signals from intelligent life. That year they searched two nearby stars, the yellow G star Tau Ceti and the orange K star Epsilon Eridani. But they didn't find any life.

They will keep looking. But radio signals indicate only intelligent life. For most of its history, Earth had only unintelligent life. Perhaps that would be easier to find.

One way to find such life is to discover a planet with lots of oxygen in its air. Oxygen is in Earth's air only because plants, trees, and other living beings put it there. In fact, if all life vanished from Earth, the oxygen would soon disappear from the air, because oxygen quickly combines with other elements. So a planet with lots of oxygen in its air is probably a planet with life.

Although we have learned much about the stars—how they are born, how they live, how they die, and how they enrich the Galaxy with life-giving elements—the stars still hold many secrets. Perhaps, as you look at the sky tonight, someone up there is looking at us, wondering what is going on around the Sun.

Glossary

A

Absolute Magnitude The amount of light a star sends into space. The absolute magnitude is the apparent magnitude a star would have if we saw it from a distance of 32.6 light-years. To compute a star's absolute magnitude, you need to know the star's apparent magnitude, its distance, and the amount of dimming caused by interstellar gas and dust between Earth and the star. See also **Apparent Magnitude**.

Albireo A beautiful double star in Cygnus made of an orange star and a blue star.

Aldebaran The brightest star in Taurus, Aldebaran is an orange K-type giant 67 light-years from Earth.

Algeiba A beautiful double golden star in Leo.

Algol The best-known eclipsing binary, Algol fades when one star passes in front of the other.

Alpha Centauri The closest star system to the Sun. Alpha Centauri is triple, so it actually has three stars: Alpha Centauri A, a yellow G-type main-sequence star like the Sun; Alpha Centauri B, an orange K-type main-sequence star somewhat fainter than the Sun; and Alpha Centauri C, a red M-type dwarf. The red star is slightly closer to us, at a distance of 4.24 light-years, so it is usually called Proxima Centauri; the other two stars are 4.37 light-years from the Sun. Alpha Centauri is slightly older than the Sun.

Altair A white A-type main-sequence star 16.7 light-years from Earth and the brightest star in the constellation Aquila. Altair spins so fast it has flattened itself a bit.

Apparent Magnitude The brightness of a star as it looks from Earth. A first-magnitude star looks brighter than a second-magnitude star, which looks brighter than a third-magnitude star, and so on. The faintest star the unaided eye can see is sixth magnitude. See also **Absolute Magnitude**.

Arcturus An orange K-type giant 37 light-years from Earth and the fourth brightest star in the night.

Atom A tiny part of matter. Every neutral atom has at least one proton, a particle with positive electric charge, and at least one electron, which has negative electric charge; all atoms except hydrogen-1 also contain at least one neutron, which has no charge. The number of protons in an atom is the atomic number and determines the element. For example, hydrogen is atomic number 1, so every hydrogen atom has one proton; helium is atomic number 2, so every helium atom has two protons; and carbon is atomic number 6, so every carbon atom has six protons.

B

Barnard's Star The second closest star system to the Sun, after Alpha Centauri. Barnard's Star is a single red dwarf 5.98 light-years away.

Betelgeuse A red star in the constellation Orion and the brightest red supergiant in the sky. Someday Betelgeuse will explode as a supernova.

Big Bang The explosion that gave birth to our universe. Within minutes, the big bang's heat created hydrogen, helium, and a little lithium, the three lightest elements.

Binary A double star in which the two stars go around each other, as opposed to an optical double star, in which the two stars appear together only because they happen to lie in the same direction from Earth.

Black Hole A dead collapsed star with so much gravity that nothing, not even light, can escape. The best-known example is Cygnus X-1.

Blueshift The shift to shorter, or bluer, wavelengths of a spectrum. Normally a blueshift occurs because a star is moving toward us, so its light waves get scrunched up, reducing their wavelength.

Blue Supergiant A big, bright, hot O or B star similar to Rigel in Orion.

Brown Dwarf A star born with so little mass that it fails to sustain the same hydrogen-to-helium nuclear reaction that occurs in main-sequence stars like the Sun. Brown dwarfs have less than 8 percent of the Sun's mass. When young, they glow red; as they cool, they fade to black.

Cygnus X-1

C

Canopus The second brightest star in the night, Canopus is a yellow-white F-type supergiant in the southern sky. It is 310 light-years from Earth.

Capella The sixth brightest star in the night, Capella is 43 light-years from Earth and quadruple. It has two yellow giants plus two red dwarfs.

Carbon The element that all life on Earth is based on, carbon is atomic number 6, which means each carbon atom has six protons.

Carbon Monoxide A molecule made of one carbon atom and one oxygen atom: CO. Carbon monoxide emits radio waves that are 2.6 millimeters long and marks the presence of molecular gas in space—the type that gives birth to stars.

Castor A bright star system in Gemini, 51 light-years from Earth, that has six stars: four white, two red.

Celsius A temperature scale that most people outside the United States use. On this scale, 0 degrees corresponds to freezing water and 100 degrees to boiling water. See also **Fahrenheit** and **Kelvin**.

Center of Mass The imaginary point in space around which stars in a star system move. In a double star system, the center of mass lies between the two stars and is closer to the more massive one.

Cepheid A yellow supergiant star that pulsates with a period of days, weeks, or months. As the star expands and shrinks, it brightens and fades. Cepheids obey a period-luminosity relation: the longer the period, the bigger and more luminous the star. Thus, measuring a Cepheid's period reveals its luminosity and therefore its distance. Polaris, Eta Aquilae, and Delta Cephei are all Cepheids.

Constellation An area of the sky. Also, a pattern of stars in this area of the sky. Astronomers divide the sky into 88 constellations. Examples include Orion, Taurus, and Scorpius.

Crab Nebula A supernova remnant 6,500 light-years away in Taurus. It is debris from a star our ancestors saw explode in 1054.

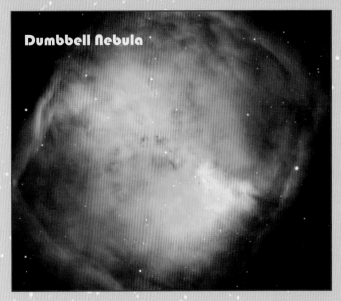

Dumbbell Nebula

Crab Nebula

Cygnus X-1 The first black hole discovered, in 1971. A blue star goes around this black hole, spilling material that gets heated so much it emits x-rays.

D

Delta Cephei The second Cepheid discovered, in 1784, which gave its name to the entire class of pulsating yellow supergiant stars. Delta Cephei pulsates every 5.3663 days.

Density Mass divided by volume. Lead has a high density; cotton has a low density. A white dwarf star has a high density; a red supergiant star has a low density.

Diameter The distance from one edge of a star through its center and to the opposite edge. The Sun's diameter is 864,900 miles, or 1,392,000 kilometers.

Double Star Two stars that appear together in the sky. In a binary, or true double star, the two stars go around each other. In an optical double, the two stars merely happen to lie in the same direction from Earth but are so far apart that they do not orbit each other.

Dumbbell Nebula A planetary nebula in the constellation Vulpecula.

Dwarf Star If not preceded by *brown* or *white*, *dwarf* means a main-sequence star—a star that makes energy by changing hydrogen into helium at its center. The term can be misleading, because some dwarfs shine very brightly; for example, Theta^1 Orionis C is the brightest star in the Orion Nebula and shines thousands of times more brightly than the Sun, yet it is a dwarf, because it is on the main sequence. The term exists because main-sequence stars are more compact than the giants and supergiants they will later become. See also **Brown Dwarf; Red Dwarf; White Dwarf**.

E

Earth Our planet. The Earth orbits the Sun once a year.

Eclipsing Binary Two stars that orbit each other so that at least one periodically blocks the light of the other, causing the system to dim as viewed from Earth. The first eclipsing binary discovered was Algol.

Electron A particle with negative electric charge that occurs in all neutral atoms.

Epsilon Eridani A young orange K-type dwarf star, $10^1/_2$ light-years from the Sun.

Epsilon Indi A triple star in the southern sky, 11.8 light-years from Earth. Epsilon Indi has an orange K-type dwarf and two brown dwarfs.

Eta Aquilae The first Cepheid discovered, in 1784, Eta Aquilae pulsates once every 7.1767 days.

Event Horizon A black hole's point of no return: once you fall through the event horizon, you can't get out, because the gravity is just too strong.

Extrasolar Planet A planet beyond our solar system.

F

Fahrenheit A temperature scale that most Americans use. In Fahrenheit, water freezes at 32 degrees and boils at 212 degrees. The Fahrenheit scale is convenient because at ordinary temperatures it does not use negative numbers or triple digits. Also, a degree in Fahrenheit is smaller than a degree in Celsius or Kelvin, so a two-digit number describes temperature more precisely than Celsius does.

To convert Celsius into Fahrenheit, multiply the Celsius temperature by 1.8 and then add 32; to convert Fahrenheit into Celsius, subtract 32 from the Fahrenheit temperature, then divide by 1.8. To convert Kelvin into Fahrenheit, multiply the Kelvin temperature by 1.8, then subtract 460; to convert Fahrenheit into Kelvin, add 460 to the Fahrenheit temperature, then divide by 1.8.

51 Pegasi The first Sunlike star found to have an extrasolar planet, 51 Pegasi happens to be 51 light-years from Earth in the constellation Pegasus. The planet is a giant world so close to the star that an orbit takes just 4¼ days.

Fomalhaut A white A-type main-sequence star located 25 light-years from Earth. Fomalhaut is the same distance from Earth as Vega.

Galaxy

G

Galaxy A large gathering of stars, gas, dust, and dark matter held together by gravity. Our Galaxy is the Milky Way; it includes every star the unaided eye can see.

Gamma Crucis The closest M-type red giant star to Earth, Gamma Crucis is 89 light-years from the Sun in Crux, the Southern Cross.

Gas Pressure The outward force of a gas. In a star, gas pressure opposes the inward pull of the star's own gravity and thus prevents the star from collapsing. The warmer the gas, the greater the gas pressure. Stars are hot, so their gas pressure is strong and holds them up against the force of their own gravity. Pressure from photons also helps out, especially in the most luminous stars—see **Photon**.

Gliese 229 A red dwarf star located 19 light-years from Earth. The first definite brown dwarf, Gliese 229 B, was found orbiting this star.

Globular Cluster A star cluster that typically has hundreds of thousands of stars packed close to one another. Compare with **Open Cluster**.

Gravitational Redshift The shift of a star's spectrum to longer, or redder, wavelengths as light struggles to leave the star. The gravitational redshift is large for white dwarf stars, which are dense and have high surface gravities.

Gravity The force that attracts one mass to another mass. The greater the masses and the closer they are to each other, the stronger the gravitational attraction.

H

H I Neutral atomic hydrogen gas: hydrogen in which every atom has both a proton and an electron. H I emits radio waves that are 8 inches, or 21 centimeters, long.

H II Ionized hydrogen gas: hydrogen in which electrons have been torn away from protons by the extreme ultraviolet light of hot stars. H II regions, such as the Lagoon Nebula, usually glow reddish.

Helium The second lightest and second most common element in the universe. Most helium in the universe arose in the big bang, but stars also produce the element. The helium on Earth, however, comes from the decay of radioactive elements such as thorium and uranium.

Helix Nebula A planetary nebula in the constellation Aquarius.

Hertzsprung-Russell (H-R) Diagram A graph that plots the luminosities and temperatures of stars. On the H-R diagram, stars fall into three groups: the main sequence, which includes the Sun; red giants and supergiants; and white dwarfs.

Horsehead Nebula An eerie dark nebula that looks like a chess piece in the constellation Orion. It is giving birth to new stars.

Horsehead Nebula

Hyades The nearest star cluster to Earth, the Hyades is located 150 light-years away in the constellation Taurus. It is an open star cluster that is 600 million years old.

Hydrogen The lightest and most common element in the universe. Hydrogen is atomic number 1, so every hydrogen atom has one proton. The Sun and most other stars are made mostly of hydrogen. Hydrogen was forged in the big bang.

I

Infrared Radiation that travels at the speed of light but has a somewhat longer wavelength than we can see. Dust in space emits infrared light.

Interstellar Medium The space between the stars.

Ionized Hydrogen Interstellar hydrogen in which electrons have been torn from protons by the extreme ultraviolet light of hot stars. Regions of ionized hydrogen, such as the Lagoon Nebula, usually glow reddish.

Lagoon Nebula

Iron Element with atomic number 26, which means every iron atom has twenty-six protons. Iron is made by supernova explosions.

J

Jupiter The largest planet that goes around the Sun. Jupiter has 318 times more mass than the Earth but only 1/1000 the mass of the Sun.

K

Kelvin A temperature scale that astronomers use. In Kelvin, absolute zero—the coldest possible temperature—is 0. So unlike Fahrenheit and Celsius, Kelvin has no negative numbers. A degree in Kelvin is the same size as a degree in Celsius and 1.8 times the size of a degree in Fahrenheit. Water freezes at 273 Kelvin and boils at 373 Kelvin.

To convert Celsius into Kelvin, add 273 to the Celsius temperature; to convert Kelvin into Celsius, subtract 273 from the Kelvin temperature. To convert from Kelvin to Fahrenheit, multiply the Kelvin temperature by 1.8, then subtract 460; to convert from Fahrenheit to Kelvin, add 460 to the Fahrenheit temperature, then divide by 1.8. The Sun's surface temperature is about 5,780 Kelvin, 5,510 degrees Celsius, and 9,940 degrees Fahrenheit.

L

Lagoon Nebula A cloud of gas and dust visible to the unaided eye in the constellation Sagittarius. The Lagoon Nebula has many newborn stars.

Lalande 21185 A red dwarf star in the constellation Ursa Major, 8.3 light-years from the Sun, and the fourth nearest star system to the Sun.

Light-Year The distance light travels in a tropical year: 5,878,499,800,000 miles, or 9,460,528,400,000 kilometers. (A tropical year is the time from the start of spring to the start of the next spring. It is about twenty minutes shorter than a sidereal year, which is the time Earth takes to orbit the Sun.) One light-year is 63,239.7 times greater than the distance from the Sun to the Earth, so the number of Sun-Earth distances in a light-year nearly equals the number of inches in a mile (63,360). The nearest star system to the Sun is Alpha Centauri, which is a bit more than 4 light-years away. This tells us that we see Alpha Centauri as it looked 4 years ago.

Luminosity The amount of light a star sends into space.

Molecular Gas in the Eagle Nebula

M

M4 One of the closest globular star clusters to Earth. M4 is 7,000 light-years distant and appears near the red star Antares in the constellation Scorpius.

Magnitude A measure of brightness. Apparent magnitude says how bright a star *looks*; absolute magnitude says how much light a star sends into space. See also **Absolute Magnitude; Apparent Magnitude**.

Main-Sequence Star A star that makes energy by changing hydrogen into helium at its center. The more mass a main-sequence star has, the bigger it is, the brighter it is, the hotter it is, the bluer it is, and the sooner it dies. The Sun is a main-sequence star; so are most other stars.

Mass The amount of material in an object. Gravity comes from mass, so the more mass something has, the more gravity it exerts.

Methane A molecule with one carbon atom and four hydrogen atoms: CH_4. Methane gets torn apart by high temperatures, so the only stars with methane are very cool: brown dwarfs.

Milky Way Our Galaxy, an enormous barred spiral made of hundreds of billions of stars. The Milky Way is roughly a trillion times more massive than the Sun. Every star you see with the unaided eye belongs to the Milky Way, and every one, including the Sun, goes around the center of the Galaxy.

Mira A pulsating red giant star in the constellation Cetus, 300 light-years away. As the star expands and shrinks, it brightens and fades. The star's pulsation period changes over time but is around 330 days.

Molecular Hydrogen Hydrogen that is made of two hydrogen atoms joined together: H_2. The densest interstellar gas is made of molecular hydrogen and thus may be giving birth to new stars.

Molecule A substance made of two or more atoms joined together. For example, water is a molecule: each water molecule contains two hydrogen atoms and one oxygen atom, H_2O. Other molecules are molecular hydrogen (two hydrogen atoms: H_2), carbon monoxide (one carbon atom and one oxygen atom: CO), and methane (one carbon atom and four hydrogen atoms: CH_4).

Moving Group Stars that travel through space together. The best-known example is the Ursa Major moving group, which includes the five central stars of the Big Dipper. Stars in a moving group probably once belonged to a star cluster that broke apart.

N

Nebula A cloud of gas and dust in space. The plural is *nebulae*. Some nebulae, such as the Orion Nebula, have newborn stars; other nebulae, such as the Ring Nebula, are planetary nebulae—glowing shells cast off by dying red giant stars; and other nebulae, such as the Crab Nebula, are supernova remnants, the wreckage of exploded stars.

Neutral Atomic Hydrogen Hydrogen gas in which every hydrogen atom has both a proton and an electron. Neutral atomic hydrogen gas emits radio waves that are 8 inches, or 21 centimeters, long. It is also called H I.

Neutron A neutral particle—with neither positive nor negative electric charge—that appears in all atoms except the most common one, hydrogen-1.

Neutron Star A small, extremely dense dead star, a neutron star arises during a supernova explosion. Pulsars are fast-spinning neutron stars.

Nitrogen The element that fills 78 percent of Earth's air. It is atomic number 7, which means each nitrogen atom has seven protons. Most nitrogen in the universe arose in stars that did not explode. The element entered the Galaxy via planetary nebulae.

Nova An explosion on a white dwarf in a double star system that does not destroy either star. The plural is *novae*.

Nuclear Reaction A collision between parts of atoms that can create new elements and enormous amounts of energy. Most stars, including the Sun and all other main-sequence stars, make energy through nuclear reactions. However, T Tauri stars, brown dwarfs, and white dwarfs have few if any nuclear reactions.

Nucleosynthesis The creation of new elements from old ones. Stars engage in nucleosynthesis when they change one element into another.

O

Omega Centauri The Milky Way's most luminous globular star cluster, located in the constellation Centaurus. Omega Centauri emits a million times more light than the Sun.

Omicron^2 Eridani A triple star located 16.3 light-years from Earth. One star in the system—Omicron^2 Eridani B—was the first white dwarf discovered.

Open Cluster A loose, spread-out cluster of stars. Most open star clusters in the Milky Way are much younger than the Sun. Examples include the Pleiades and the Hyades, both in the constellation Taurus. Compare with **Globular Cluster**.

Orbit The path a star, planet, or moon follows through space.

Orion Nebula

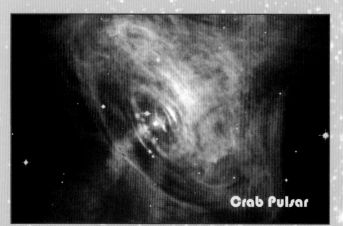

Crab Pulsar

Orion Nebula The best-known interstellar cloud of gas and dust, the Orion Nebula has thousands of newborn stars. It is visible to the unaided eye, south of Orion's belt. Recent studies indicate the Orion Nebula is 1,350 light-years from Earth, a couple hundred light-years closer than astronomers once thought.

Oxygen The element we breathe. It is atomic number 8, which means every oxygen atom has eight protons. Oxygen was made in massive stars, like Rigel, Deneb, Antares, and Betelgeuse, then cast into space when the stars exploded. Oxygen is the third most common element in the universe, after hydrogen and helium.

P

Parallax The small shift in a star's position when we view the star from slightly different angles during the year as the Earth circles the Sun. The closer a star is to Earth, the larger its parallax; thus, measuring the parallax reveals the star's distance.

Photon A particle of light. Photons exert pressure. Extremely luminous stars have so many photons streaming out of them that photon pressure is the main outward force that holds the star up against the inward-pulling force of gravity. In contrast, in most stars, including the Sun, gas pressure is the main outward force that holds the star up against gravity—see also **Gas Pressure**.

Planet An object that is much less massive than a star and that emits no light but merely reflects the light of stars around it. The Earth is a planet that goes around the Sun.

Planetary Nebula A red giant star's cast-off gas, set aglow by the star's hot, newly exposed core. Examples include the Ring Nebula, the Dumbbell Nebula, and the Helix Nebula.

Pleiades

Pleiades A young open star cluster 435 light-years from Earth in the constellation Taurus. This star cluster is a stunning sight through binoculars.

Polaris The North Star. Polaris is the brightest and nearest Cepheid. It is at the end of the Little Dipper's handle.

Pre-Main-Sequence Star A young star that is slowly shrinking, making heat and light mainly through gravity rather than nuclear reactions. The best-known pre-main-sequence star is T Tauri.

Procyon A yellow-white F-type star in Canis Minor and one of the closest stars to the Sun. Procyon is 11.4 light-years away. A white dwarf star orbits it.

Proton A particle with positive electric charge that exists in the centers of all atoms. The number of protons in the atom always equals the atomic number: thus, hydrogen, which has one proton, is atomic number 1; helium, which has two protons, is atomic number 2; oxygen, which has eight protons, is atomic number 8; and so on.

Proxima Centauri The nearest individual star to the Sun, Proxima Centauri is the faintest member of the triple Alpha Centauri star system. Proxima Centauri is a red dwarf star. It is 4.24 light-years from the Sun, slightly closer to us than Alpha Centauri A and B are.

PSR B1257+12 A pulsar in the constellation Virgo. In 1991, this pulsar was the first star found beyond the Sun to have planets.

Pulsar A fast-spinning neutron star that emits radio waves in a beam. If oriented in the right way, every time the pulsar spins, its beam hits the Earth, so observers see a pulse of radio waves.

Q

Quadruple Star A star system with four stars. The brightest quadruple star is Capella, which has two yellow stars and two red stars. Another famous quadruple star is Epsilon Lyrae. If you look at Epsilon Lyrae through binoculars, you see two stars; if you look at it through a telescope, you see that each of those stars is double, giving a total of four stars altogether.

R

Radio Waves Radiation that travels at the speed of light but has a much longer wavelength. As a result, our eyes can't see radio waves. Neutral atomic hydrogen gas in space gives off radio waves that are 8 inches, or 21 centimeters, long.

Radius The distance from the center of a star to its surface. The radius is always half the diameter. The Sun's radius is 432,500 miles, or 696,000 kilometers.

Red Dwarf A faint, cool main-sequence star of spectral type M or L. Three fourths of all stars are red dwarfs. The nearest red dwarf to the Sun is also the nearest star to the Sun: Proxima Centauri, 4.24 light-years away.

Red Giant A large, aging, luminous cool star. Billions of years from now, the Sun will swell into a red giant. Red giants eventually form planetary nebulae and become white dwarfs.

Redshift The shift to longer, or redder, wavelengths of a star's spectrum. A redshift arises when a star moves away from us, stretching the light waves so they have longer wavelengths. However, strong gravity can also cause a redshift—see **Gravitational Redshift**.

Red Supergiant An enormous, luminous cool star. The brightest in the sky are Betelgeuse, in the constellation Orion, and Antares, in the constellation Scorpius. Someday these stars will explode as supernovae.

Regulus The brightest star in the constellation Leo. Regulus is a blue B-type main-sequence star 79 light-years from Earth. Regulus spins so fast it has flattened itself a bit.

Rigel The brightest star in the constellation Orion. A blue B-type supergiant, Rigel is the most luminous star within a thousand light-years of the Earth.

Ring Nebula A beautiful planetary nebula in the constellation Lyra.

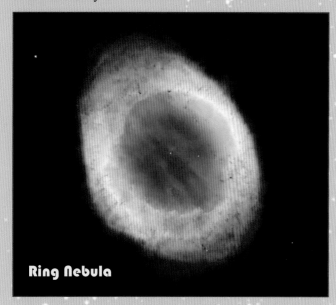

Ring Nebula

r-Process The formation of heavy elements when neutrons rapidly hit iron nuclei. The r-process probably occurs in the supernova explosions of massive stars. It made many of the elements that are heavier than iron, including most gold, silver, platinum, and iodine. See also **s-Process**.

S

Sirius The brightest star in the night. Sirius is 8.6 light-years from Earth, twice as far as Alpha Centauri, and consists of two stars: a bright white A-type main-sequence star and a dim but even hotter white dwarf. The two stars orbit each other every 50 years.
61 Cygni A double orange K-type dwarf, 61 Cygni is 11.4 light-years from the Sun. In 1838, this star made history when astronomer Friedrich Wilhelm Bessel for the first time succeeded in measuring the parallax of a star other than the Sun.
Solar Mass The amount of material in the Sun, which is 1,989,100,000,000,000,000,000,000,000,000 kilograms, or 4,385,200,000,000,000,000,000,000,000,000 pounds. The Sun has 332,946 times more mass than the Earth.
Spectral Type A classification of a star based on its spectrum. From hot and blue to cool and red, the seven main spectral types are O (blue), B (also blue), A (white), F (yellow-white), G (yellow, like the Sun), K (orange), and M (red). Astronomers have added two spectral types—L and T—for cooler stars.
Spectrum The arrangement of light by order of color, or wavelength. A rainbow is a spectrum of sunlight, because it displays sunlight in order of color, from purple to red. By studying the spectrum of a star, astronomers can deduce its surface temperature and composition. The plural of *spectrum* is *spectra*.
Spica The brightest star in the constellation Virgo, Spica is a blue star of spectral type B.
s-Process The creation of heavy elements in a red giant or supergiant star from the slow bombardment of iron nuclei by neutrons. Because neutrons are neutral—they have no electric charge—the iron nuclei, which have positive charge, do not repel them. The s-process made most of the strontium, yttrium, zirconium, niobium, molybdenum, tin, barium, lanthanum, cerium, thallium, and lead in the universe. See also **r-Process**.
Star A hot, glowing object in space that makes its own heat and light. A star shines because it is hot.

Stars in the Milky Way

Star Cluster A large group of stars held close to one another by their gravity. An open star cluster is a loose gathering that typically has hundreds of stars. Examples include the Hyades and the Pleiades. In contrast, a typical globular cluster has hundreds of thousands of stars close to one another. Examples of globular clusters include M4 in the constellation Scorpius, M13 in the constellation Hercules, and Omega Centauri.
Star System All the stars that orbit one another because gravity holds them close together. A single star system, like the Sun, has just one star; a double star system, like Sirius, has two stars; a triple star system, like Alpha Centauri, has three stars; and so on.
Stefan-Boltzmann Law This rule says that the amount of light emitted per square inch of a star's surface depends strongly on its surface temperature. In particular, to compute the output of light per square inch relative to the Sun, multiply the star's surface temperature four times. For example, if a star is twice as hot as the Sun, then every square inch of the star's surface emits 16 times more light, because $2 \times 2 \times 2 \times 2 = 16$.
Sun The star Earth orbits. The Sun is a main-sequence star, which means it generates energy because its center changes hydrogen into helium. The Sun is yellow and spectral type G. It is 4.6 billion years old, its diameter is 864,900 miles, and its mass is 332,946 times the Earth's. On average, the Earth is 93 million miles from the Sun.
Supergiant A large star that emits roughly 10,000 times more light than the Sun. Rigel is a blue supergiant, Deneb is a white supergiant, Canopus is a yellow-white supergiant, and Antares and Betelgeuse are red supergiants.

Supernova An explosion that destroys a star. The plural is *supernovae*. Supernovae occur when massive stars—those born with more than eight solar masses—explode and become neutron stars or black holes. However, some supernovae occur when white dwarf stars receive material from another star. The latter explosions are called type Ia.

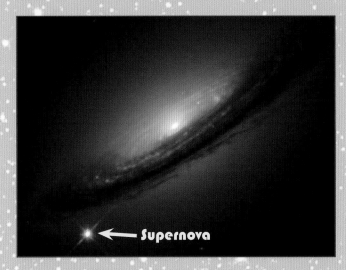

Supernova Remnant The debris from and swept up by an exploded star. The best-known supernova remnant is the Crab Nebula in the constellation Taurus. Other examples include Cassiopeia A and the Veil Nebula.

Supernova Remnant Cassiopeia A

T

Tau Ceti The nearest single yellow G-type star to the Sun. It is 11.9 light-years from Earth. Tau Ceti has fewer heavy elements than the Sun.
Temperature A measure of how hot something is. The hottest stars are blue, followed by white, then yellow-white, then yellow, then orange, and finally red.
Tide A distortion an object suffers because the side of the object facing a mass feels a stronger gravitational pull than the side facing away.

Triple Star A star system with three stars. The nearest triple star is Alpha Centauri.
T Tauri A young, pre-main-sequence star in the constellation Taurus. It resembles our Sun 4 1/2 billion years ago.
Type Ia Supernova An exploding white dwarf star. Its spectrum shows no hydrogen. Other supernovae come from dying high-mass stars.

U

Ultraviolet Radiation whose wavelength is slightly shorter than the human eye can see. Ultraviolet photons travel at the speed of light but have more energy than photons of visible light.
Universe Everything that exists: all the planets, stars, and galaxies.

V

Variable Star A star whose brightness changes. Examples include T Tauri, a pre-main-sequence star; Algol, an eclipsing binary; Polaris, a Cepheid; and Mira, a pulsating red giant.
Vega The fifth brightest star in the night. Vega is a white A-type main-sequence star located 25 light-years from Earth in the constellation Lyra.
Veil Nebula An old supernova remnant in the constellation Cygnus.

W

Water A molecule made of two hydrogen atoms and one oxygen atom: H_2O. All life on Earth needs water.
White Dwarf A small, dim, dense, dying star. A white dwarf has used up its nuclear fuel and is now cooling and fading. The nearest white dwarf is Sirius B, 8.6 light-years from the Sun. In 7.8 billion years, the Sun will become a white dwarf.
Wolf 359 The third closest star system to the Sun, Wolf 359 is a dim red dwarf 7.8 light-years away. German astronomer Max Wolf reported the star's discovery in 1918.

X

X-Rays High-energy radiation that travels at the speed of light. X-rays can arise from very hot material, such as the gas that falls into a black hole.

Y

Yellow Supergiant A star whose temperature is similar to the Sun's but emits roughly 10,000 times more light. Examples include Polaris and Canopus. Some yellow supergiants pulsate and are Cepheids.

Z

Zzzzzzzzzz What astronomers do on a cloudy night.

Index

A

Absolute magnitude, 16–17, 18–19, 24
Abundances of elements, 55, 57
    and extraterrestrial life, 61
    in globular clusters, 52
    of hydrogen, 8, 55, 57
    in open clusters, 52
Achernar, 18, 21
Adams, Walter, 33
Air
    density, 8
    nitrogen and oxygen in, 54, 61
    blocks x-rays, 45
Albireo, 18, 19, 46, 47
Aldebaran, 19, 24
Algeiba, 19, 47
Algol, 18, 47
Alkaid, 18, 21
Alpha Centauri
    distance, 46, 47
    on H-R diagram, 19
    life around, 61
    mass, 21
    planets, 58, 61
    spectral types, 20, 21, 46, 61
    as triple star, 46–47
Altair, 18, 20, 21
Aluminum, 56–57
America, 16, 33, 41
Andromeda Galaxy, 41
Antares, 19, 34, 37
Apparent magnitude, 16
Arcturus, 19, 24
Aristotle, 40
Atomic hydrogen gas, 8, 10
Atomic numbers, 56
Austria, 7

B

Baade, Walter, 42
Barnard's Star, 19, 20, 21
Bell, Jocelyn, 42
Bessel, Friedrich Wilhelm, 32–33
Betelgeuse, 17, 19, 34, 35, 37, 40
Big bang, 54, 55, 56–57
Big Dipper, 52
Binaries, eclipsing, 47
Binary stars, 20–21, 22, 23, 46–47, 61
Birth of planets, 61
Birth of stars, 14–15
Black holes, 5, 44–45
Blueshift, 59
Blue supergiant stars, 34, 35, 37
Boltzmann, Ludwig, 7
Brown dwarfs, 22–23

C

Calcium, 5, 17, 55, 56–57
California, 42
Capella, 19, 24
Carbon
    abundance, 57
    burning, 49, 56
    creation, 25, 54, 56–57
    and life, 54
    made by massive stars, 54, 56–57
    in planetary nebulae, 26, 54, 57
    made by red giants, 25, 54, 56–57
    in white dwarfs, 48, 49
Carbon monoxide, interstellar, 12
Cassiopeia A, 38
Castor, 47
Cat's Eye Nebula, 31
Center of mass, 47
Cepheids, 40–41
Cepheus, 41
Cetus, 25
Chi Draconis, 18, 19, 21
Clark, Alvan Graham, 33
Cone Nebula, 12, 13
Copper, 55, 56–57
Crab Nebula, 36–37, 43
Crab pulsar, 43
Cygnus, 34, 37, 45, 47
Cygnus X-1, 44, 45

D

Delta Cephei, 19, 41
Deneb, 18, 34
Denmark, 16
Diameter
    of brown dwarfs, 23
    defined, 7
    largest, 5, 34
    of neutron stars, 5, 43
    of red giants, 24
    of red supergiants, 34
    of Sun, 7, 25
    of white dwarfs, 32, 33
Double stars, 20–21, 22, 23, 46–47, 61
Dumbbell Nebula, 26, 29
Dust in space, 8, 10, 14–15, 38, 39

E

Eagle Nebula, 12
Earth
    age, 21
    composition, 55
    diameter, 7
    future, 24, 25, 33
    gravity, 14
    life, 61
    photograph, 60
    our planet, 5, 6
    tides, 45
Eclipsing binaries, 47
Electrons, 10, 11, 33, 42, 43, 49
Elements, 56–57
    interstellar, 8, 10–12
    origin, 5, 54–57
    in planetary nebulae, 26
    in spectra, 17
Elkin, William, 6
Elliptical orbits, 47
England, 26, 40, 41, 42, 50
Epsilon Eridani, 19, 20, 21, 61
Epsilon Indi, 19, 20, 21, 23
Eskimo Nebula, 30
Eta Aquilae, 18, 41
Europa, 61
Event horizon, 45
Extrasolar planets, 23, 58–59, 61
Extraterrestrial life, 58, 60–61

F

"Failed" stars, 22–23
51 Pegasi, 58–59
First-magnitude stars, defined, 16
Fomalhaut, 18, 21
Formation of planets, 61
Formation of stars, 14–15
Frail, Dale, 58

G

Galaxies
    Andromeda, 41
    Cepheids in, 40–41
    measuring distances to, 40–41, 49
    Milky Way. *See* Milky Way Galaxy
    NGC 3370, 40, 41
    NGC 4526, 49
    supernovae in, 42, 48, 49
Gamma Crucis, 19, 24
Gamma Virginis, 18, 21
Gas, interstellar, 8–13, 14, 50
Gas pressure, 14, 15, 25, 34
Gemini, 47
Germany, 26
Giant molecular clouds, 50
Giant stars, 17, 24–25, 26, 52, 55, 56
Gill, David, 6
Gliese 229, 19, 22–23
Globular clusters, 52, 53
Gold, 55, 56–57
Goodricke, John, 41
Gravitational redshift, 33
Gravity
    of black holes, 44–45
    in brown dwarfs, 22, 23
    defined, 14
    and double stars, 20–21, 33, 47, 61
    of extrasolar planets, 58, 59
    makes heat and light, 14, 15, 22
    in massive stars, 34, 37
    of neutron stars, 43
    of red giants, 25
    shifts light to the red, 33
    and star birth, 14–15
    in star clusters, 50, 52
    in supergiants, 34, 37
    and star death, 37
    and tides, 45
    of white dwarfs, 33, 49
Greece, 40

H

H I, 8, 10
H II, 10–11
Harvard University, 32
Helium
    abundance, 57
    from big bang, 54, 55, 56–57
    burning to carbon and oxygen, 25, 37, 56
    from hydrogen burning, 15, 20, 21, 25, 34, 37
    in stellar spectra, 17
    in Sun, 15, 55
Helix Nebula, 28
Herschel, William, 26
Hertzsprung, Ejnar, 16
Hertzsprung-Russell (H-R) diagram, 16–19
Horsehead Nebula, 8, 10, 12
Hot Jupiters, 59

H-R diagram, 16–19
Hyades cluster, 50, 52
Hydrogen
    abundance, 8, 55, 57
    atomic, 8, 10
    from big bang, 54, 55, 56–57
    burning to helium, 15, 20, 21, 25, 34, 37
    H I, 8, 10
    H II, 10–11
    interstellar, 8, 10–12
    ionized, 10–11
    molecular, 11–12, 50
    in nebulae, 8, 10–12
    neutral atomic, 8, 10
    radio waves from, 10
    in Sun, 15, 55
    not in type Ia supernovae, 48

I

Infrared radiation, 14–15, 25, 38
Intelligence, extraterrestrial, 60–61
Interstellar medium, 8–13
    atomic hydrogen (H I), 8, 10
    carbon monoxide, 12
    composition, 8, 10–12
    density, 8
    dust, 8, 10, 14–15, 38, 39
    hydrogen, 8, 10–12
    ionized hydrogen (H II), 10–11
    mass of, in Milky Way, 8
    molecular, 11–12, 50
    and star formation, 14
    temperature, 14
Iodine, 55, 56–57
Ionized hydrogen, 10–11
Iron, 5, 37, 49, 52, 54, 55, 56–57, 61

J

Jupiter, 5, 6, 22, 23, 34, 58, 59, 61

K

Kelvin, defined, 7

L

Lagoon Nebula, 8, 11, 12
Lalande 21185, 19, 20
Lead, 55, 56–57
Leavitt, Henrietta, 41
Leo, 20, 47
Life in space, 60–61
    and elements, 5, 54, 55
    and planets, 58, 59, 61
Light speed, 6, 44, 45
Light-year
    and absolute magnitude, 16
    defined, 6
Lithium, 54
Luminosity
    and absolute magnitude, 16–17, 18–19, 24
    calculation of, 6, 7, 16–17
    defined, 6
    on H-R diagram, 16–17, 18–19
Lyra, 26

M

M4 (globular star cluster), 52
M80 (globular star cluster), 53

Magnesium, 56–57
Magnitude, 16–17, 18–19, 24
Main-sequence stars, 20–21
    defined, 15, 17, 20
    and element creation, 56
    on H-R diagram, 17, 18–19, 20
    and life, 61
    massive, 34
    and red giants, 24
    in star clusters, 52
Mars, 5, 24, 25, 58, 61
Mass transfer, 47, 48–49
Mayor, Michel, 58–59
Mercury, 5, 24, 25, 47, 59, 61
Meteorites, 21
Methane in brown dwarfs, 22–23
Milky Way Galaxy
    elemental abundances, 52, 54, 55, 57
    last supernova in, 37
    location of interstellar gas, 10, 12
    mass of interstellar matter, 8
    number of black holes, 45
    number of globular clusters, 52
    number of open clusters, 50
    number of stars, 5, 61
    star formation rate, 14
Mira, 19, 24, 25, 40
Molecular gas, 11–12, 50
Moon, 14, 17, 34, 45
Moving group, 50, 52
Mu Columbae, 18, 21
Multiple star systems, 46–47

N

Nebulae, 8–13
    colors, 11, 26
    density, 8
    planetary, 26–31, 54, 57, 61
    supernova remnants, 36–39
    temperature, 14, 38
Neon, 56–57
Neptune, 47
Neutral atomic hydrogen gas, 8, 10
Neutrons, 42, 43, 49, 55
Neutron stars, 38, 42–43, 44, 49
NGC 3370, 40, 41
NGC 4526, 49
Nitrogen, 26, 54, 55, 56–57
North Star, 19, 41
Novae, 47
Nuclear reactions
    in brown dwarfs, 22
    defined, 15
    and element creation, 54–57
    in main-sequence stars, 20–21, 25, 34, 56
    in red giants, 25
    r-process, 55, 56
    s-process, 55, 56
    in Sun, 15, 25
    in supergiants, 37
    in supernovae, 49, 55, 56
Nucleosynthesis, 5, 54–57

O

Omega Centauri, 52
Omicron^2 Eridani, 18, 19, 32
Open star clusters, 50–52
Orange dwarf stars, 20, 21, 23, 46, 61
Orange giant stars, 24, 47, 52

Orange supergiant stars, 34, 37
Orbits, elliptical, 47
Orion, 8, 17, 34, 35
Orion Nebula, 8, 9, 12
Oval orbits, 47
Oxygen
    abundance, 55, 57
    burning, 56
    creation, 5, 25, 54, 56–57
    in Earth, 55
    in old stars, 52, 61
    and search for life, 61
    in white dwarfs, 48

P

Parallax, 6
Pavo, 22
Pegasus, 58
Perseus, 47
Phosphorus, 56–57
Photon pressure, 15, 34
Photons, 15, 26, 33
Pickering, Edward, 32
Pigott, Edward, 40–41
Planetary nebulae, 26–31, 54, 57, 61
Planets
    around brown dwarfs, 23
    defined, 5
    extrasolar, 23, 58–59, 61
Platinum, 55, 56–57
Pleiades cluster, 50–51, 52
Pluto, 6, 14
Pneumonia, 41
Polaris, 19, 41
Potassium, 56–57
Pre-main-sequence stars, 14–15
Proctor, Richard, 50
Procyon, 18, 20, 24
Procyon B, 18, 33
Protein, 54, 55
Protons, 10, 11, 15, 21, 42, 43, 55
Proxima Centauri, 19, 20, 21, 46–47
Prussia, 32
PSR B1257+12, 58
Pulsars, 42–43, 58
Pulsating stars, 24, 25, 40–41

Q

Quadruple stars, 46
Queloz, Didier, 58–59
Quintuple stars, 46
Quintuplet cluster, 52

R

Radiation pressure, 15, 34
Radio telescopes, 10, 42, 45
Radio waves
    from atomic hydrogen gas, 10
    from black hole Cygnus X-1, 45
    from carbon monoxide, 12
    from life in space, 61
    from molecular gas, 12
    from pulsars, 42, 43
Red dwarf stars, 20, 21, 22, 23, 46–47, 59
Red giant stars, 17, 24–25, 26, 33, 52, 55, 56
Redshift, 33, 59
Red supergiant stars, 17, 34–35, 37, 42, 55, 56
Regulus, 18, 20, 21, 24

Rigel, 17, 18, 34, 35, 40
Ring Nebula, 26, 27, 54
r-Process, 55, 56
Russell, Henry Norris, 16, 32

S

Satellite, x-ray, 45
Saturn, 47, 58, 59
Scorpius, 34, 52
SCR 1845-6357, 19, 22, 23
Second-magnitude stars, defined, 16
Sextuple stars, 46, 47
Silicon, 55, 56–57
Silver, 56–57
Sirius
    absolute magnitude, 16
    apparent magnitude, 16
    compared with Sun, 6, 7
    distance, 6, 16
    as double star, 20–21, 46, 47
    on H-R diagram, 16, 18
    luminosity, 6
    magnitude, 16
    mass, 20–21, 47
    orbital period, 21, 46
    parallax, 6
    photograph, 32
    spectral type, 20, 21
Sirius B, 18, 20–21, 32–33, 46, 47, 48
61 Cygni, 19, 20
Sodium, 22, 56–57
South Africa, 6
Spectral types, defined, 17, 23
Spectrum, 17, 22, 33, 48, 59
s-Process, 55, 56
Star clusters, 50–53
Stars
    absolute magnitude, 16–17, 18–19, 24
    ages, 52, 61
    apparent brightness, 6–7, 16
    apparent magnitude, 16
    biggest, 5, 34
    binary, 20–21, 22, 23, 46–47, 61
    birth, 14–15
    black holes, 5, 44–45
    blue supergiant, 34, 35, 37
    brightest at night, 6
    brown dwarf, 22–23
    central temperature, 15, 25
    Cepheid, 40–41
    closest, 5, 46–47
    colors, 17, 18–19, 20, 22, 24, 25, 33, 34, 37, 40, 47
    defined, 5
    diameters, 5, 21, 23, 24, 33, 34, 43
    distance, measuring, 6, 41
    double, 20–21, 22, 23, 46–47, 61
    eclipsing binaries, 47
    and element creation, 5, 54–57
    "failed," 22–23
    formation, 14–15
    inefficient, 15
    ionize interstellar hydrogen, 10–11
    largest, 5, 34
    least luminous, 5, 20, 21, 22–23
    lifetimes, 21, 61
    luminosities, 6, 7, 16–17, 18–19, 21
    magnitudes, 16–17, 18–19, 24
    main-sequence, 15, 17, 18–19, 20–21, 24, 34, 56, 61
    mass, 20–21, 33, 44, 47
    mass transfer, 47, 48–49
    measuring diameters, 34
    measuring masses, 20–21, 47
    most luminous, 5, 34
    nearest, 5, 46–47
    neutron, 38, 42–43, 44, 49
    newborn, 12, 14–15
    orange dwarf, 20, 21, 23, 46, 61
    orange giant, 24, 47, 52
    orange supergiant, 34, 37
    parallax, 6
    pre-main-sequence, 14–15
    pulsating, 24, 25, 40–41
    red dwarf, 20, 21, 22, 23, 46–47, 59
    red giant, 17, 24–25, 26, 52, 55, 56
    red supergiant, 17, 34–35, 37, 42, 55, 56
    small, 5, 23, 32, 33, 43
    supergiant, 17, 18–19, 34–35, 37, 40–41, 42, 55, 56
    surface temperature, 7, 15, 17, 18–19, 21, 23, 26, 33
    triple, 46–47
    T Tauri, 14–15
    variable, 24, 25, 40–41, 47
    white dwarf, 17, 18–19, 32–33, 47, 48–49, 55, 56–57
    white supergiant, 34, 37
    yellow giant, 24, 47
    yellow main-sequence, 16, 17, 20, 21, 58–59, 61
    yellow supergiant, 34, 37, 40–41
Star system, defined, 46
Stefan, Josef, 7
Stefan-Boltzmann law
    defined, 7
    and red supergiants, 34
    and white dwarfs, 33
Strong force, 15
Sulfur, 56–57
Sun
    absolute magnitude, 16
    age, 15, 21, 50, 55, 61
    apparent magnitude, 16
    average star?, 17
    brighter in future, 24, 25
    central temperature at birth, 15
    central temperature today, 25
    closest star, 5, 47
    compared with other stars, 17
    compared with Sirius, 6, 7
    composition, 55
    diameter, 7
    distance from Earth, 6
    expansion rate, 25
    faint when young, 25
    future of, 24, 25, 33
    on H-R diagram, 16, 19
    lifetime, 21
    magnitude, 16
    maximum brightness, 25
    not an average star, 17
    photon pressure, 15
    radiation pressure, 15
    as red giant, 24, 25
    spectral type, 17, 20, 21
    surface temperature, 7
    temperature at center at birth, 15
    temperature at center today, 25
    temperature at surface, 7
    as white dwarf, 33
Supergiant stars, 17, 18–19, 34–35, 37, 40–41, 42, 55, 56
Supernovae
    and element creation, 49, 54, 55, 56–57, 61
    from massive stars, 37, 42
    from white dwarfs (type Ia), 48–49, 55, 56–57
Supernova remnants, 36–39, 43
Switzerland, 58

T

Tau Ceti, 19, 20, 21, 61
Taurus, 15, 24, 37
Third-magnitude stars, defined, 16
Tides, 45
Tin, 56–57
Triple stars, 46–47
T Tauri, 14, 15
Type Ia supernovae, 48–49, 55, 56–57

U

Ultraviolet radiation
    ionizes hydrogen, 11
    in planetary nebulae, 26
    from Sun, 26
United States, 16, 33, 41
Uranium, 55, 56–57
Uranus, 21, 26, 46
Ursa Major moving group, 52

V

Van Maanen's Star, 18, 33
Variable stars
    Cepheids, 40–41
    defined, 40
    eclipsing binaries, 47
    Miras, 24, 25, 40
Vega, 16, 18, 20, 21, 24
Veil Nebula, 37, 39
Venus, 5, 24, 25, 58, 59, 61
Virgo, 58
Vitamin D, 26
Vulpecula, 42

W

Water, 59, 61
White dwarf stars, 32–33
    and element creation, 49, 55, 56–57
    on H-R diagram, 17, 18–19
    and novae, 47
    and type Ia supernovae, 48–49, 55, 56–57
White supergiant stars, 34, 37
Wolf 359, 19, 20
Wolszczan, Alex, 58

X

X-rays, 38, 43, 45

Y

Yellow giant stars, 24, 47
Yellow main-sequence stars, 16, 17, 20, 21, 58–59, 61
Yellow supergiant stars, 34, 37, 40–41

Z

Zero-magnitude stars, defined, 16
Zeta Ophiuchi, 18, 21
Zwicky, Fritz, 42